I0393798

Table of Contents

Page

Abstract iv

Acknowledgments v

List of Figures viii

List of Tables x

List of Abbreviations xi

1   Introduction 1
  1.1   The Focal Space Mission 1
  1.2   Thesis Objective 1

2   Background & Theory 4
  2.1   Overview 4
  2.2   Gravitational Lens 4
  2.3   Secondary Missions 13
  2.4   Space Missions 16
    2.4.1   Attitude Control and Propulsion 17
    2.4.2   Communications and Data-handling 19
    2.4.3   Electrical Power 23
    2.4.4   Environmental Control 27
  2.5   Conclusion 32

3   Analysis 33
  3.1   Introduction 33
  3.2   Launch Vehicle 33
  3.3   Communications and Data Handling 37
    3.3.1   Shannon Limit 40
    3.3.2   Signal Modulation 42
    3.3.3   Antennas 44
    3.3.4   Autonomy and Lifespan 45
  3.4   Power 46
  3.5   Thruster 47
    3.5.1   Introduction 47
    3.5.2 ΔV 47
    3.5.3   Payload Mass Fraction 52

     3.5.4   Power 53

     3.5.5   Gravity Assist 53

  3.6   Attitude Control 58

  3.7   Environmental Control 62

  3.8   Conclusion 64

4   Design 65

  4.1   Design Flow 65

  4.2   Launch Vehicle 67

  4.3   Spacecraft 68

     4.3.1   Payload 68

     4.3.2   Communications and Data Handling 68

     4.3.3   Power 69

     4.3.4   Propulsion 73

     4.3.5   Attitude 75

         4.3.5.1   Attitude Determination 75

         4.3.5.2   Attitude Control 76

     4.3.6   Environmental Control 79

  4.4   Model 81

  4.5   Performance Analysis 87

     4.5.1   Mass 87

     4.5.2   Power 88

     4.5.3   Trajectory Analysis 89

         4.5.3.1   Pre-Jupiter Phase 90

         4.5.3.2   Jupiter Phase 91

         4.5.3.3   Post-Jupiter Phase 92

         4.5.3.4   Total 92

     4.5.4   Communications 93

     4.5.5   Cost 94

5   Conclusions and Recommendations 96

  5.1   Introduction 96

  5.2   Conclusions 96

  5.3   Recommendations 97

List of Figures

Figure   Page

2.1   Geometry of the Sun's Gravitational Lens. Note: figure is not drawn to scale. . .   5

2.2   Critical Frequency as a Function of Impact Parameter   8

2.3   Focusing Conditions   9

2.4   Locus of Focal Minima 11

2.5   Minimum Impact Parameter as a Function of Frequency 12

2.6   Venetia Burney Student Dust Counter   [1] 14

2.7   Juno's Magnetometer Locations   [2] 15

2.8   Pioneer X and XI Configuration [3]   17

2.9   Dawn Trajectory [4] 20

2.10 New Horizons Data Rate   [5] 23

2.11 Solar Irradiance 24

2.12 Juno Solar Panels   [6] 25

2.13 Theoretical Decay of Plutonium-238 27

2.14 Ulysses Spacecraft   [7] 29

3.1   Selected Launch Vehicle Performance   [8] 35

3.2   Effect of Specific Impulse on ΔV 36

3.3   Space Loss   39

3.4   Shannon Limit   41

3.5   Constant EIRP Curves (W) 42

3.6   Mission Data Rate   44

3.7   Delta-V Required for Excess Velocity   49

3.8   Delta-V for Various Thrust Levels   51

3.9   Delta-V for Various Specific Impulses   52

3.10 Time Between Launch Opportunities   [8] [9] 54

3.11 Voyager I Gravity Assist Velocity Relative to the Sun [10] 57

3.12 Pointing Accuracy Geometry 58

4.1 Design Flow Diagram 66

4.2 Payload Fairing [11] 67

4.3 Payload and Communications Architecture. Note: The primary payload serves as a radio telescope and HGA. 70

4.4 General Purpose Heat Source RTG [12] 70

4.5 Power Source Trade 71

4.6 NEXT Thruster System [13] 74

4.7 Attitude Control Loop 79

4.8 Internal Configuration 82

4.9 Stowed Configuration 83

4.10 Zoomed Out View 84

4.11 RTG View 85

4.12 Zoomed In View 86

4.13 Spacecraft with Boost Phase 87

4.14 RTG Power Level and Spacecraft Power Draw 89

4.15 Pre-Jupiter Trajectory (Distance - AU)(Angle - degrees) 90

4.16 Pre-Jupiter Velocity Profile 91

4.17 Post-Jupiter Velocity Profile 92

4.18 Post-Jupiter Trajectory (Distance - AU)(Angle - degrees) 93

4.19 Mission Data Rate 94

List of Tables

Table   Page

1.1   Thesis Objectives Summary 3

2.1   Juno Magnetometer Table   [2] 15

2.2   Attitude Control History 18

2.3   DSN Allocated Frequency Band (MHz) [14] 23

2.4   RTG Evolution 26

3.1   Modulation Schemes 43

3.2   Typical Thermal Ranges for Spacecraft Components   [15] 63

4.1   Payload Performance   [15] 69

4.2   Power Performance 73

4.3   NEXT Sizing and Power [13] 73

4.4   NEXT Component List [13] 75

4.5   Attitude Determination Mass and Power 76

4.6   FSM Mass Moments of Inertia 77

4.7   FSM Slew Rate 78

4.8   Propellant Tank Design 79

4.9   Spacecraft Heat Balance 81

4.10   Spacecraft Dry Mass Budget 88

4.11   Spacecraft Power Consumption 88

List of Abbreviations

Abbreviation		Page
FSM	Focal Space Mission	1
CMB	Cosmic Microwave Background	1
AU	Astronomical Unit	1
TAU	Thousand Astronomical Unit	13
VBSDC	Venetia Burney Student Dust Counter	14
FGM	Flux Gate Magnetometer	16
PEPSSI	Pluto Energetic Particle Spectrometer	16
NASA	National Aeronautics and Space Administration	16
LGA	Low Gain Antenna	21
HGA	High Gain Antenna	21
MGA	Medium Gain Antenna	21
TWTA	Travelling Wave Tube Amplifiers	22
DSN	Deep Space Network	23
RTG	Radioisotope Thermoelectric Generators	26
EOL	End of Life	26
ASRG	Advanced Stirling Radioisotope Generator	26
GPHS	General Purpose Heat Source	27
RHU	Radioisotope Heater Unit	30
BER	Bit Error Rate	42
HPBW	Half Power Beamwidth	45
COTS	Commercial-off-the-Shelf	46
TRL	Technology Readiness Level	46
JGA	Jupiter Gravity Assist	53
SGA	Saturn Gravity Assist	53

UGA   Uranus Gravity Assist 53

NGA   Neptune Gravity Assist 53

JSGA   Jupiter and Saturn Gravity Assist 53

JUGA   Jupiter and Uranus Gravity Assist 53

JNGA   Jupiter and Neptune Gravity Assist 53

SoGA   Solar Gravity Assist 53

VSGA   Venus and Solar Gravity Assist 53

NEXT   NASA Evolutionary Xenon Thruster 73

GRAVITATIONAL LENS: DEEP SPACE PROBE DESIGN

1 Introduction

1.1 The Focal Space Mission

The Focal Space Mission (FSM) revolves around the concept of gravitational lensing proposed by both Albert Einstein and Orest Khvolson [16] . Gravitational lensing occurs when electromagnetic waves travel in the vicinity of massive objects. The gravitational field of the object bends, or lenses, the waves and causes them to converge to a focal point where they can be resolved [16]. Any significantly sized astronomical object exhibits this phenomenon, which was confirmed in 1979 by Walsh, Carswell, and Weymann [17]. Furthermore, this lensing effect greatly magnifies objects from a background source. Utilizing the sun as a gravitational lens provides an unmatched opportunity to observe both neighboring and distant astronomical phenomena in exceptional detail.

The primary mission of the FSM will be to use radio telescopy in concert with the sun's gravitational lens to image nearby solar systems and/or other pertinent astronomical objects, e.g. the Cosmic Microwave Background (CMB) radiation.

1.2 Thesis Objective

In *Deep Space Flight and Communications* Claudio Maccone uses the mass, radius, mean density, and Schwarzschild radius of the sun to find its minimum focal distance. His result was 548 Astronomical Units (AU) or 13.86 times the Sun-to-Pluto distance [18]. With that, Dr. Maccone gives an in-depth review of the physics necessary to complete this mission; however, there has not been a fully developed system design of a spacecraft to move this idea forward. Therefore, the two aims of this thesis are to provide a FSM mission analysis and FSM spacecraft design. However, unlike other notable innovative interstellar precursor designs, e.g. Project Prometheus, the FSM design will consist solely of commercially available technology (COTS). This will allow current and future investigators to understand the limits of current deep space capabilities so that recommendations can be made for going forward. Finally, it is important to note the design done in this study is limited in scope owing to the fact that it is a masters thesis and does not have the requisite resources for a full-up spacecraft design. Consequently, there is no fault-tree analysis and certain subsystems were explored in more depth than others.

The mission analysis will explore the feasibility of traveling to the focal point of the gravitational lens and provide estimates for what is required to perform the mission. This includes examining the ΔV required, the trip time, and the trajectory of the spacecraft. On the other hand, the spacecraft design will include a system trade between subsystems, a spacecraft model, and performance measures. The spacecraft design will include a system trade between subsystems, a spacecraft model, and performance measures. Therefore the five deliverables are a trajectory analysis, subsystem trade, spacecraft model, spacecraft performance analysis and launch vehicle analysis.

More specifically, the trajectory analysis will look into the ΔV required, trip time, and trajectory. The subsystem trade will look at the attitude control system, propulsion system, electrical power system, environmental control system, and communications system, and of these subsystems the power, propulsion and communications subsystems will receive the most scrutiny because they are the greatest challenges to mission success. The two remaining systems will loosely adopt solutions used on other comparable missions. Next, the spacecraft model will display both the internal spacecraft layout and the external spacecraft layout. The external spacecraft layout will include models of the spacecraft in the stored and deployed configuration. Following that, the spacecraft performance analysis will address whether the mission can be completed with the assembled spacecraft. The launch vehicle analysis will explore things from mass and volume constraints to the ΔV it provides. The final and unifying goal of the study is to understand whether there is sufficient technological maturity to complete an interstellar precursor mission with COTS. Concluding this section, Table 1.1 outlines the results and where each can be found in the document while Table 4.1 outlines the mission payloads.

Table 1.1: Thesis Objectives Summary

Thesis Objectives	Objective Breakout	Result	Section Number
Trajectory Analysis	ΔV	51.79 km/sec	4.5.1.4
	Trip Time	108.7 years	4.5.1.4
	Trajectory	Spiral	4.5.1
Subsystem Trade	Attitude Control System	Star Tracker IMU Att. Thruster	4.3.5
	Propulsion System	NEXT	4.3.4
	Power System	GPHS	4.3.3
	Environmental Control System	MLI RHU Radiators/w Louvers	4.3.6
	Communication System	12-m HGA 2-m MGA	4.3.2
Model	External		4.4
	Internal		4.4
Performance Analysis	Data Rate	Variable	4.5.4
	Dry Mass	2345.93 kg	4.5.2
	Power Consumption	188.75 W	4.5.3
	Cost	$3-5 Billion (FY11 US$)	4.5.5

2 Background & Theory

2.1 Overview

The first step in a novel scientific study is a comprehensive literature review. Thus, this chapter is present to positively assert the unique nature of the FSM mission by analyzing NASA's previous deep space missions, where deep space is defined to be greater than 5 AU, and to present a basis from which to design the FSM. Accompanying the analysis of past space missions is a rigorous treatment of the science and history of gravitational lensing, since it is of critical importance to the subsequent work presented, along with a discussion of secondary missions and payloads.

2.2 Gravitational Lens

It is only recently that the phenomenon of gravitational lensing has come to be understood. As stated earlier, it was posited in the early 20th century and not verified until 1979. Einstein introduced the world to the physics of the gravitational lens in his seminal *Science* journal article titled *Lens-like Action of a Star by the Deviation of Light in the Gravitational Field* [16]. The geometry of the sun's gravitational lens is displayed in Figure 2.1, which is not to scale. It must be noted that *A* does not need to be a star and indeed it can be any observable phenomenon. In the figure, α_0 corresponds to the deflection angle, *x* to the distance from the line that intersects stars *A* and *B*, R_0 to the radius of the sun, and *D* corresponds to the radial distance from the sun to its focus.

In his article, Einstein put forth that at the focal point *D* , of star *B*, the star *A* does not appear to be a star at all but a "luminous circle" with angular radius β, where

$$\beta = \sqrt{\alpha_0 \frac{R_0}{D}} \quad [16] \tag{2.1}$$

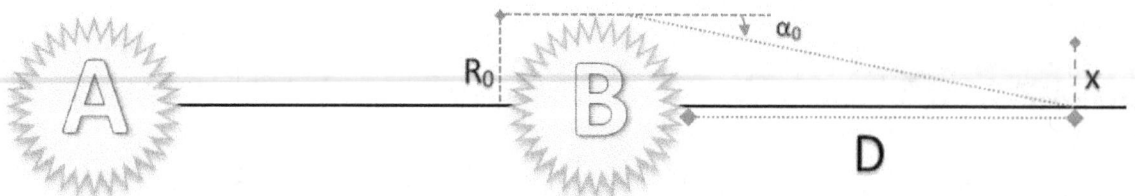

Figure 2.1: Geometry of the Sun's Gravitational Lens. Note: figure is not drawn to scale.

and

$$\alpha_0 = \frac{4GM_{sun}}{c^2 r} \quad [18]$$

(2.2)

With the introduction of the first equations, it is important to note that all equations listed in this thesis use base International System units unless otherwise noted.

In the Equation 2.2, G represents the universal gravitational constant, M_{sun} represents the mass of the sun, c represents the speed of light, and r represents the radial distance at which the light passes over the sun. Furthermore, anything in this luminous circle will have an increased apparent brightness due to the gravitational field of B. However, this is limited by x, the distance from the line that intersects the center of star A and B, because as the observer moves farther away from the line of intersection the magnification decreases [16]. This ratio of magnification, q, is dictated by the expression

$$q = \frac{l}{x} \frac{1 + \dfrac{x^2}{2l^2}}{\sqrt{1 + \dfrac{x^2}{4l^2}}} \quad [16]$$

(2.3)

where

$$l = \sqrt{\alpha_0 D R_0} \quad [16]$$

(2.4)

Looking at Equation 2.3 the reader will immediately notice that to have a considerable level of magnification, $q \gg 1$, x must be very small compared to l. Thus, the

5

value $\frac{x^2}{l^2}$ can be excluded from the calculation for the cases germane to this thesis yielding the equation [16]

$$q = \frac{l}{x} \quad [16] \tag{2.5}$$

To maximize the magnification x must be minimized and l must be maximized. In his article, Einstein noted the interesting quality that as the displacement from the line of intersection approaches zero the magnification goes to infinity [16]. The first quantity x can only be affected by the attitude and trajectory of the spacecraft. The second quantity l is a function of two variables, one being the distance from the focusing star D and the second being the deflection angle α_0. As the distance along the line of intersection increases the magnification increases proportionally to $\sqrt{D}$. This means that the gravitational focus is a line extending outward to infinity from the minimum focal point [18]. So maximizing this value is also dependent upon the trajectory of the spacecraft. On the other hand, the deflection angle α_0 has a definite maximum. It occurs at the minimum possible radius. In the case of the Sun this minimal radius is just the radius of the sun R_0. Therefore, electromagnetic waves deflected at R_0 will maximize α_0, l, and as a result q. However, at this point and with the information provided thus far, utilizing the trigonometry of Figure 2.1 the angle α can be related to the distance from the focusing star, D and the radius of the sun R_0

$$\tan \alpha = \frac{R_0}{D} \quad [18] \tag{2.6}$$

which through the small angle approximation and substitution becomes

$$D = \frac{R_0^2}{2r_g} \quad [18] \tag{2.7}$$

where the Schwarzschild Radius r_g is

$$r_g = \frac{2GM_{sun}}{c^2} \quad [18] \tag{2.8}$$

This yields a minimum focal distance of 542 *AU* [18]. Yet, caution should be exercised with this result because due to the sun's chaotic corona the focal distance for specific frequencies can be pushed out hundreds of astronomical units or even completely eliminated.

The sun's corona, according to the Baumbach-Allen Model, is composed primarily of three sources: K-corona, F-corona, and E-corona [18]. Utilizing this, the most accurate model of the corona, Anderson and Turyshev produced Equation 2.9 for the deflection angle due to the corona.

$$\alpha_{corona} = \left(\frac{\nu}{6.36\,MHz}\right)^2 \left[2952\left(\frac{R_0}{b}\right)^{16} + 228\left(\frac{R_0}{b}\right)^6 + 1.1\left(\frac{R_0}{b}\right)^2\right] \quad [19] \qquad (2.9)$$

where *b* is the impact parameter, or radius from the center of the sun to the point of deflection, and ν is the frequency of the wave to be observed. So the total deflection becomes

$$\alpha_{total} = \alpha - \alpha_{corona} \quad [18] \qquad (2.10)$$

This implies that the total deflection angle is a function of the impact parameter as well as frequency. This brings about a very important result–the critical frequency. If the frequency of the wave is below this critical frequency for a given impact parameter then the focus disappears [18]. The critical frequency is given by the equation

$$\nu_{critical}(b) = \sqrt{\frac{(6.36\,MHz)^2 R_0}{2r_g}\left(2952\left(\frac{R_0}{b}\right)^{15} + 228\left(\frac{R_0}{b}\right)^5 + 1.1\left(\frac{R_0}{b}\right)\right)} \quad [18] \qquad (2.11)$$

Plotting this equation (Eq. 2.11), by varying the impact parameter *b* from 1 to 15, produces Figure 2.2. Also, with the equation for the critical frequency a new equation for the location of the gravitational focus emerges. Equation 2.12 clearly demonstrates that if the frequency is less than the critical frequency of its impact parameter then the focus

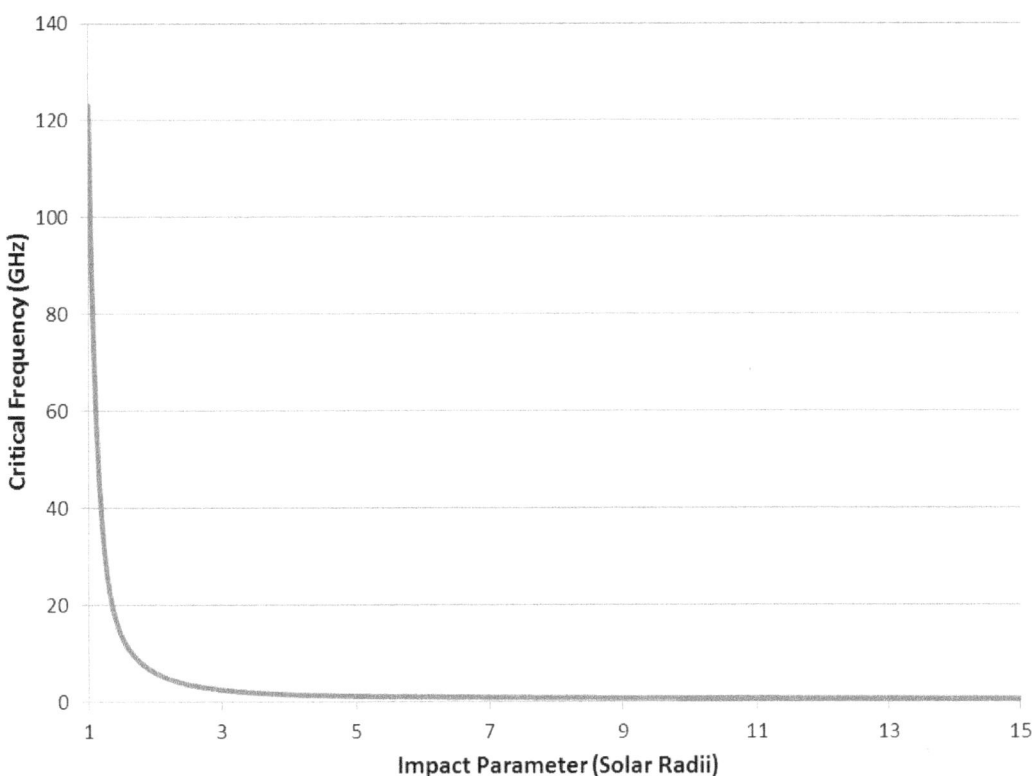

Figure 2.2: Critical Frequency as a Function of Impact Parameter

disappears because its point of intersection with the focal axis is negative.

$$D_{total} \approx \frac{542 \left(\frac{b}{R_0} \right)^2}{1 - \frac{v^2_{critical}(b)}{v^2}} \quad [18] \qquad (2.12)$$

Noting that the focus existence is dependent upon the impact parameter and frequency Dr. Maccone developed focusing conditions based on Baumbach-Allen model. To do this Dr. Maccone split the Baumbach-Allen model into its three constituent pieces and found focusing conditions for each. The E-Corona corresponds to impact parameters in the range of 1 to 1.3 solar radii [18]. The K-Corona corresponds to impact parameters in the range of 1.3 to 3 solar radii [18]. Finally, the F-Corona corresponds to impact parameters from three solar radii to infinity [18]. Looking at the equations for coronal deflection

8

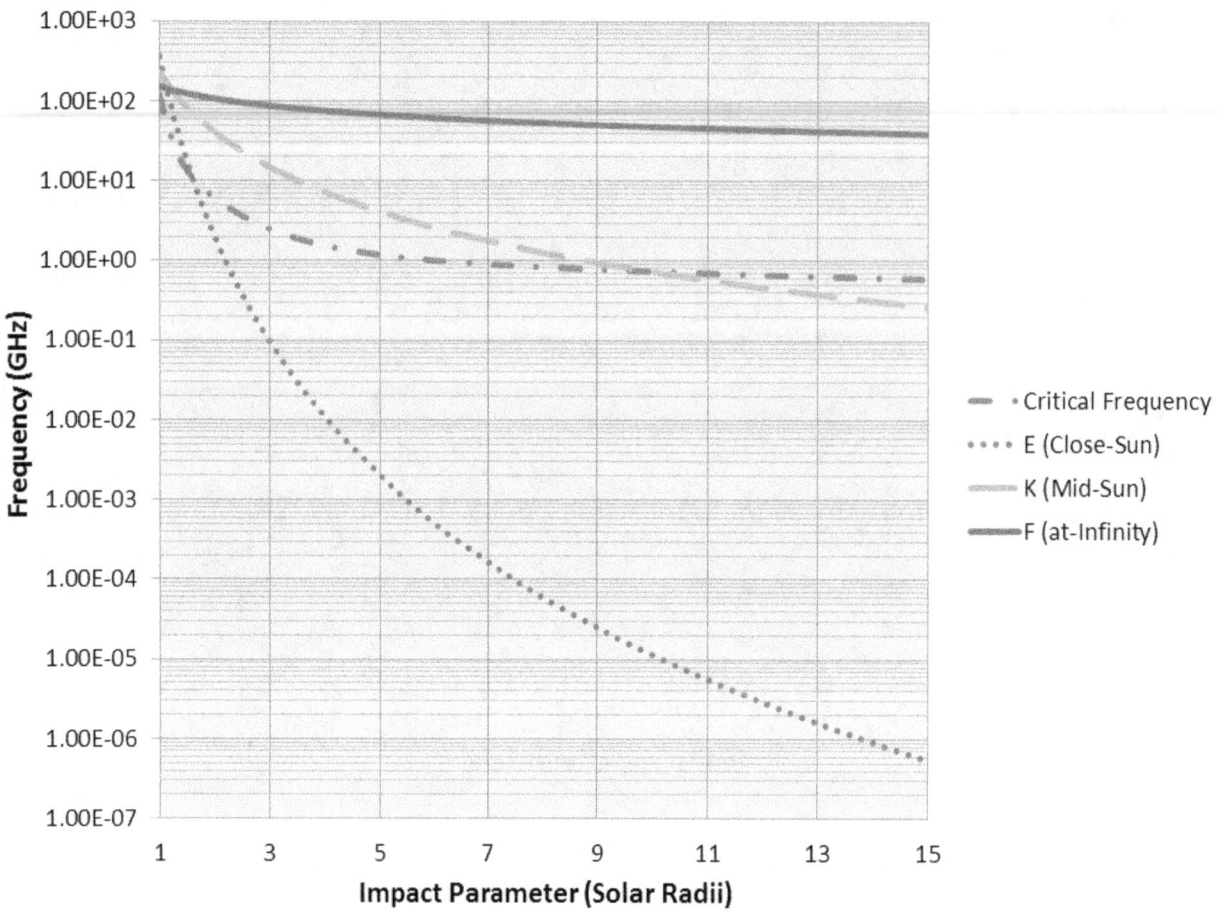

Figure 2.3: Focusing Conditions

(Eq. 2.9) the E-Corona, K-Corona, and F-Corona are represented by the three terms in brackets and in that order. After this separation Dr. Maccone noted that the observing frequency, which is a function of the impact parameter, for each piece of the corona must be greater than its respective critical frequency [18]. By plotting the equations for the observing frequency of each of the coronal constituents versus the critical frequency the focusing conditions can be found. This is shown in Figure 2.3.

Each focusing condition is located by finding the intersection between the critical frequency and coronal constituent. For the E-Corona the intersection point occurs at (1.588,11.07). This implies that focusing occurs for impact parameters less than 1.588

solar radii and frequencies above 11.07 GHz in this zone. The K-Corona's intersection point is located at $(10, .75)$. Finally, the F-Corona has focusing for all observing frequencies since there is no intersection.

Aggregating all of the information presented thus far yields the most important equation of the section–Equation 2.13. From Einstein it was apparent that the gravitational focus is not a point but a line extending to infinity from the minimum focal distance [16]. Referring back to Equation 2.7 the sun sans corona, or naked sun, focal distance minimum occurs when the waves graze the naked sun at R_0. The introduction of the Baumbach-Allen corona model resulted in Equation 2.12. This new equation necessitated the exploration of focusing conditions. To explore these conditions the corona model was broken down into its three constituent pieces. From this analysis it was clear that even with the introduction of the corona the logic gleaned from Equation 2.7 still holds meaning the minimum focal distance will occur for waves travelling closest to the naked sun in the E-Corona. Thus by going back and isolating the E-Corona component, producing the equation for the straight ray path, minimizing that equation and finally simplifying Equation 2.13 is produced. Equation 2.13 is the locus of focal distance minima. Figure 2.4 shows the simple parabolic nature of the locus. The E-Corona focusing conditions state that the waves must travel less than 1.58 solar radii and that is why Figure 2.4 terminates at that point. With that b_{min} is plotted as a function of frequency in Figure 2.5 which has a lower limit defined by the E-corona focusing condition and an upper limit defined by the minimum impact parameter of one.

$$D_{min}(b_{min}) = \frac{17}{30} \frac{b_{min}^2}{r_g} \quad [18] \qquad (2.13)$$

where

$$b_{min}(v) = \frac{17^{\frac{1}{15}} (v_{critical}^{\frac{2}{15}}(R_0)) R_0}{2^{\frac{1}{15}} v^{\frac{2}{15}}} \quad [18] \qquad (2.14)$$

10

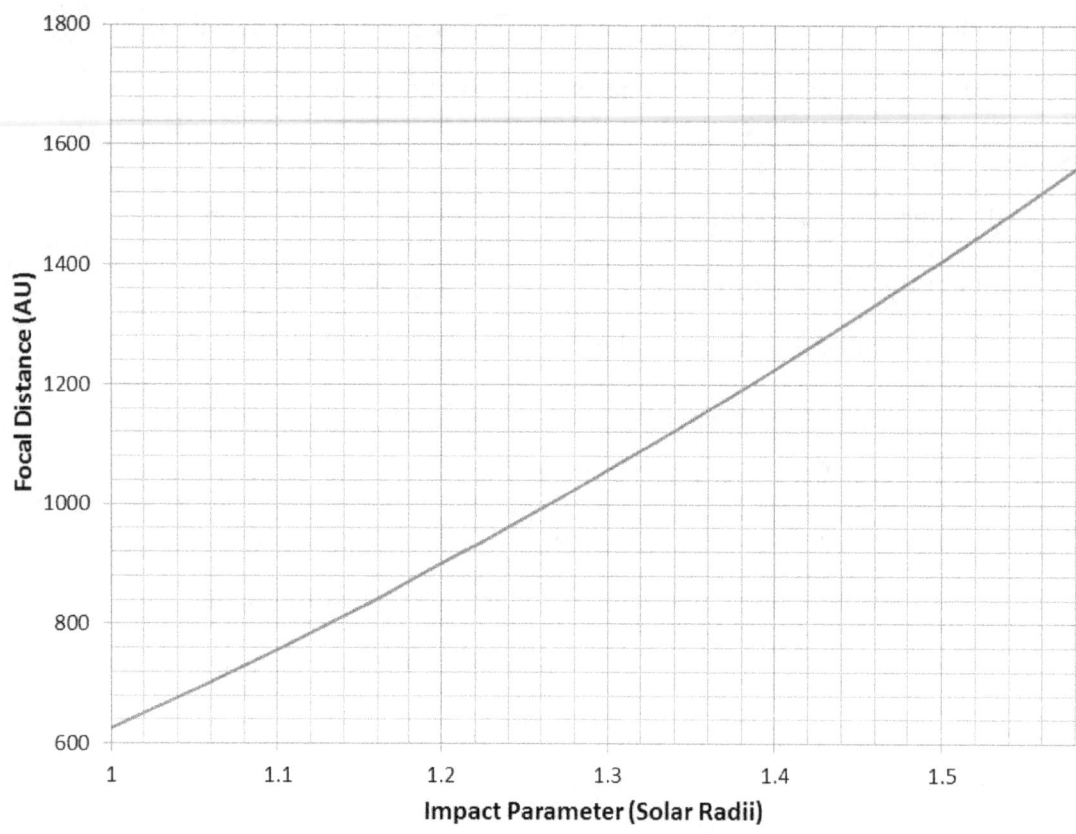

Figure 2.4: Locus of Focal Minima

With these two figures the absolute minimum focal distance is found to reside at ~625 AU and at a frequency of ~355 GHz . To find other focal distances the reader may choose a desired frequency in the range of 11.07 to 455 GHz and use Figure 2.5 to find the corresponding impact parameter. Then using that impact parameter go to Figure 2.4 and locate the corresponding minimum focal distance.

To utilize the gravitational lens Dr. Maccone proposed the use of a 12 meter parabolic radio telescope, which will serve as the primary payload in this design [18]. This payload will allow the spacecraft to measure the variations of the sun's focus to characterize it and to take images of opportunity using radio telescopy. Though optical telescopes have been used to investigate gravitational lensing on Earth optical telescopes

11

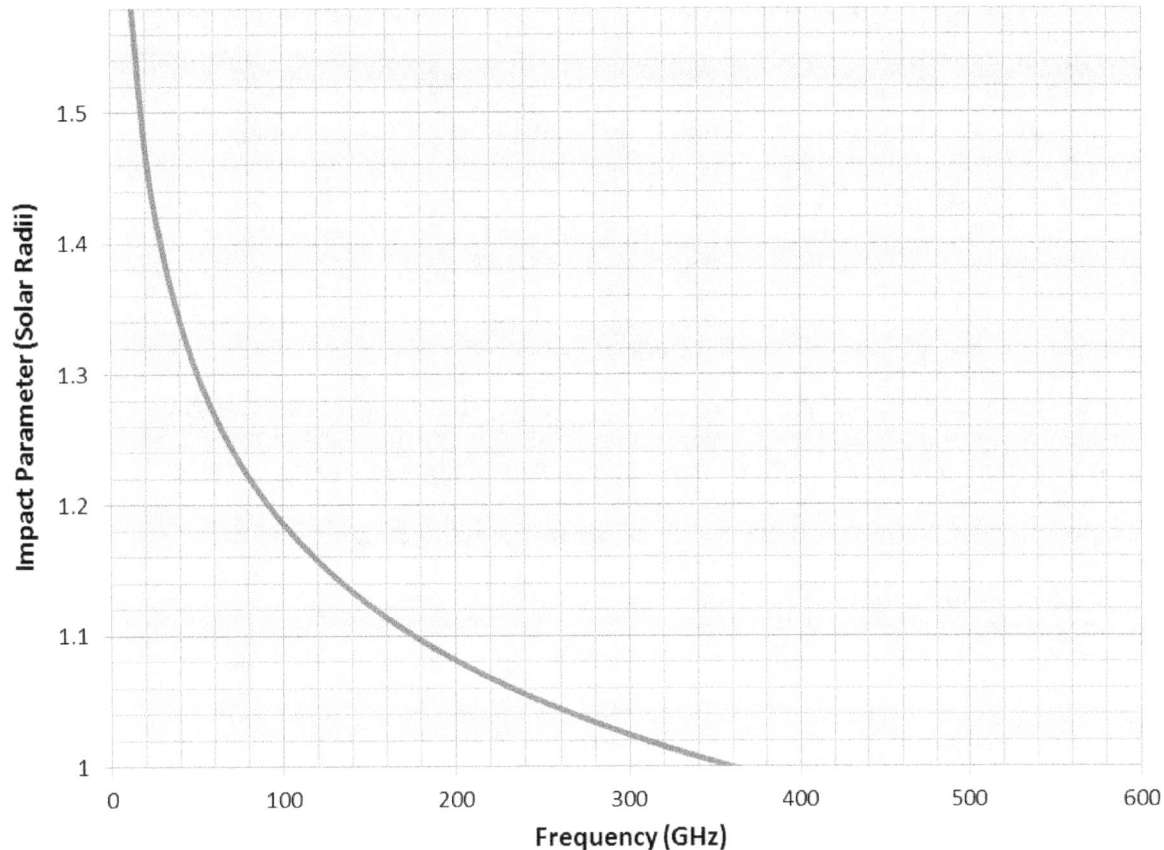

Figure 2.5: Minimum Impact Parameter as a Function of Frequency

were not investigated in this study because optical frequencies, in the range of 400 - 750 THz, are not present in the close sun approximation. As a result, their focusing distance would be beyond 1000 AU.

The resultant gain from the radio telescope plus the sun would be

$$G_{Total} = \frac{16\pi^4 GM_{Sun}r_{antenna}^2 v^3}{c^5} \quad [18] \tag{2.15}$$

where G is the gravitational constant, M_{Sun} is the mass of the sun, $r_{antenna}$ is the radius of the antenna, c is the speed of light, and v is the frequency being observed. This equation motivates the mission.

It is clear from Equation 2.15 that increasing the observed frequency has a dramatic effect on the total gain. However, the opposite is true for the image size. The image size is defined as the distance x in Figure 2.1 where the gain falls by 6 dB [18]. Mathematically, the image size is illustrated by the equation

$$r_{ImageSize} = \frac{c^2 \sqrt{D}}{2\pi^2 \sqrt{GM_{sun}}v} \quad [18] \tag{2.16}$$

where D is the distance of the aperture from the sun and all other variables are the same as they appear in Equation 2.15. Though the frequency makes the image smaller, as the spacecraft travels away from the sun the radius of the image will get larger and the magnification/resolution will also improve.

The physics of gravitational lensing is well established and has been verified. Nonetheless, the exact location of the sun's gravitational focus is still merely an approximation due to the calculation's reliance on computational and theoretical models. But what is known is that the focus will lie somewhere between 550 and 800 AU. Consequently, that will be the primary mission of the spacecraft–determine the exact location gravitational lens focus, utilize it, improve the model of the lens, and study astronomical phenomena.

2.3 Secondary Missions

Since utilizing the gravitational lens requires the spacecraft to travel out to such great distances, it provides the opportunity for the spacecraft to explore the interstellar medium in-situ. As a result, exploring the interstellar medium and heliosphere constitute important secondary missions.

NASA's Thousand Astronomical Unit (TAU) has already examined what would be important and scientifically viable to study in-situ. Based on the conclusions in the TAU NASA study, three secondary missions were derived.

The first mission is to examine dust particles encountered in the heliosphere and compare them to those in the interstellar medium. The mass, speed, direction, and composition are the principal characteristics of concern. These values will allow for improved models of the distribution of dust particles and inform scientists on the origin of the particles (interstellar or from within the solar system) [20]. Six spacecraft have previously flown dust detectors within the solar system [21]. The most recent to do so was New Horizons. The Venetia Burney Student Dust Counter (VBSDC), which is shown in Figure 2.6, was flown on New Horizons to measure dust levels in the inner solar system and Kuiper Belt. It can measure dust particles with masses from 10^{-12} to 10^{-9} kg and radii in the range of 1 to 10 μm [21]. Since the VBSDC would also achieve the FSM secondary mission objectives its specifications will be used for the design. The VBSDC has a mass of 1.6 kg, dimensions of 45.72 cm x 30.48 cm, and an average power of 5.1 W [21].

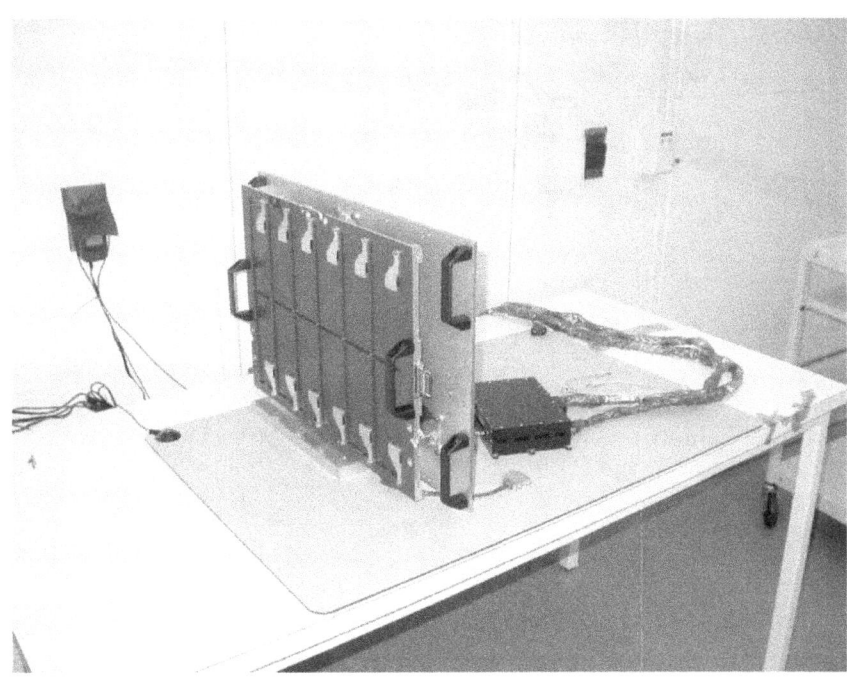

Figure 2.6: Venetia Burney Student Dust Counter   [1]

The second mission is to map the magnetic field as a function of heliocentric distance [20]. This will help physicists create more accurate models for the magnetic field of the heliosphere. It will also allow the magnetic instabilities at the heliopause to be studied in-situ [20]. Concurrently, the magnetic field mapping outside of the heliosphere, in the interstellar medium, will provide useful information that will enable physicists to construct models for the origin and generation of the galactic magnetic field [20]. Recently, Juno has flown magnetometers to map the magnetic field of Jupiter. These state of the art magnetometers could also map the heliosphere's magnetic field as a function of heliocentric distance.

Figure 2.7 displays the two magnetometers flown on the Juno mission while Table 2.1 displays their performance.

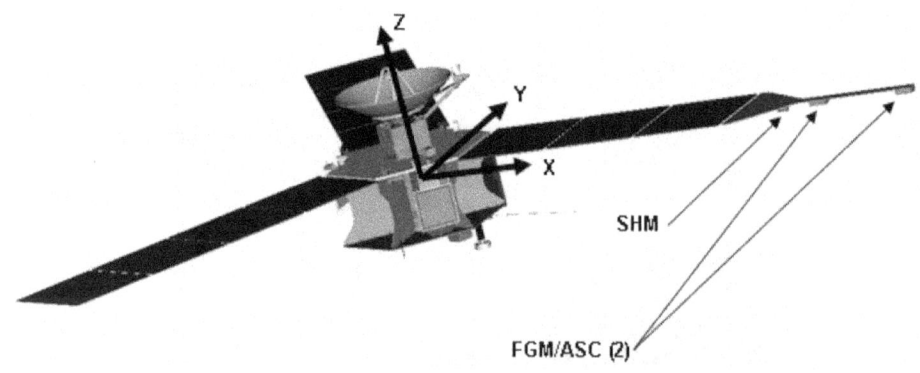

Figure 2.7: Juno's Magnetometer Locations   [2]

Table 2.1: Juno Magnetometer Table   [2]

		Juno Magnetometers	
		Flux Gate Magnetometer	Scalar Helium Magnetometer
Range	(Gauss)	$2 \times 10^{-6} - 12$	.1—12
Accuracy	(%)	.05	.002%
Power	(W)	11.3	6.5
Mass	(kg)	15.25	9.08

15

Notice where the magnetometers are placed in Figure 2.7. In missions where the objective is to use magnetometers for characterizing the magnetic field, like Juno and MagSat (which characterized the magnetic field of the Earth), the magnetometers are always placed on booms to avoid registering interference from the rest of the spacecraft bus [22]. Also, it is important to note that while the power and mass of the Flux Gate Magnetometer (FGM) are low they are, in actuality, even lower because Juno's FGM possessed an advanced star compass that will not be necessary for the heliospheric mapping. Nonetheless, the numbers for the Flux Gate Magnetometer and the Scalar Helium Magnetometer in Figure 2.1 will be used for the FSM design.

The final mission is to detect neutral and ionized particles in the interstellar medium and heliopause. This will allow scientists to determine the initial energy distribution of the interstellar medium as well as understand how the solar system interacts with the galactic environment in the outer heliosphere [20]. The Pluto Energetic Particle Spectrometer Science Investigation (PEPSSI) on New Horizons could accomplish this mission. The PEPSSI can measure ions in the range of 1 keV to 1 MeV in a 120 degree by 12 degree beam [23]. It can determine the composition and spectrum of neutral particles, ions, and electrons [23]. The PEPSSI has a mass of 1.5 kg and a power requirement of 2.5 W.

2.4 Space Missions

In 1958 the newly formed National Aeronautics and Space Administration, NASA, set out on the ambitious mission of placing a satellite into a heliocentric orbit. Two months after the Soviet Union accomplished this lofty goal with Luna-1 NASA achieved Earth escape with Pioneer IV. With their newfound understanding of how to achieve a heliocentric orbit NASA pushed their goals higher. One of the results of this was Pioneer X and XI which would explore Jupiter and Saturn.

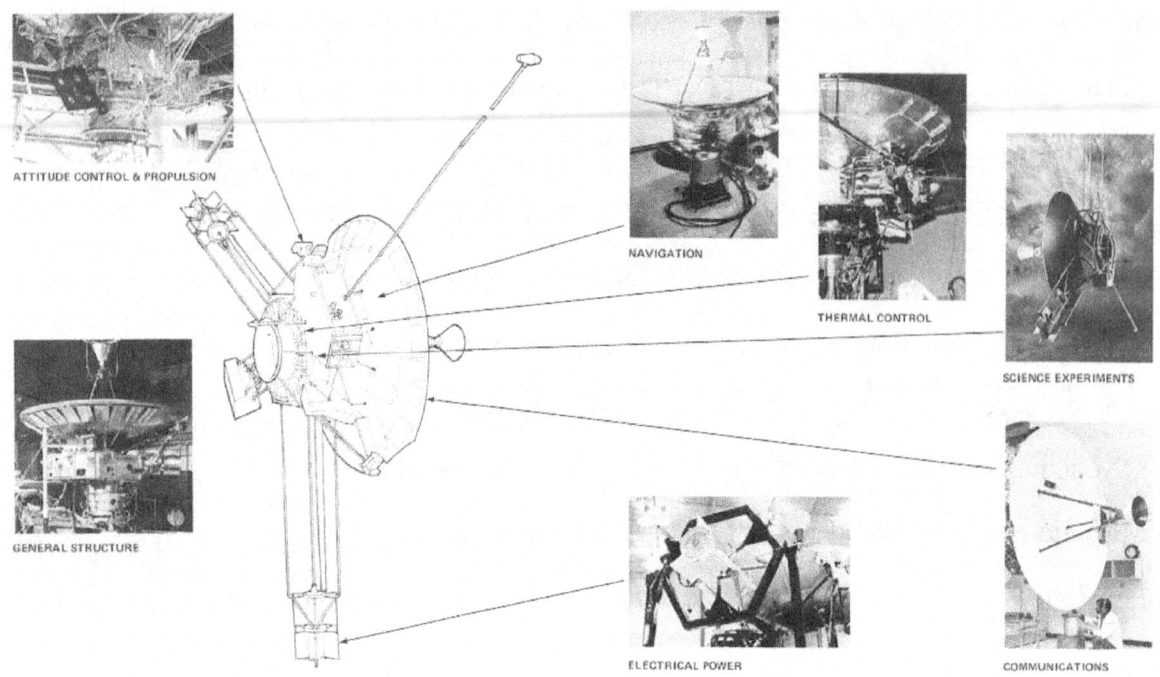

Figure 2.8: Pioneer X and XI Configuration [3]

From the Pioneer missions of the 1960s to the Juno mission launched in 2011 bold interplanetary probes have laid the foundation for the FSM. Due to the diligent work of the mission engineers invaluable lessons were learned on how to develop and implement all of the primary subsystems including attitude control and propulsion, communications and data-handling, electrical power, and environmental control. This section is meant to be an overview of current deep space capabilities. For this overview I centered my analysis on the most recent deep space probes and touched on other relevant missions where appropriate.

2.4.1   Attitude Control and Propulsion.   Like all spacecraft systems the attitude control subsystem is driven by mission requirements. Since every mission varies the sensor and actuator complements vary also.

Table 2.2 shows a complete history of sensor and actuator complements for deep space probes. Furthermore, it demonstrates the differing approaches to attitude control. For example, looking at the most recent missions–Juno and New Horizons–Juno is spin stabilized while New Horizons has multiple attitude modes. Juno's primary missions involve surveying characteristics of Jupiter and since it will be in a Jovian orbit spin stabilization is adequate for the duration. On the other hand, New Horizons will not enter into orbit around Pluto so it requires three axis stabilization to acquire specific targets on Pluto, and spin stabilization for other portions of its mission.

Though the modes vary, there is a commonality between how stabilization is maintained. There is a clear trend in both the sensor and actuator suites of the deep space probes listed in Table 2.2. In regard to the sensor complement, the standard approach is to use star trackers, sun sensors, and inertial reference units to gauge the attitude of the spacecraft. The aptly named star trackers and sun sensors gauge the position of the spacecraft based on the location of a star and the sun respectively. The inertial reference units are made up of four gyroscopes. Three of the gyroscopes are orthogonal and the fourth is skewed so that it can compensate for any gyroscope that malfunctions [24]. The inertial reference units, sometimes called inertial measurement units, measure the spacecraft's orientation as well as velocity

To maintain and correct to the proper attitude spacecraft primarily employ MMH/N_2O_4 thrusters. This is due to the high efficiency, moderate cost, and storability of the fuel. On top of their attitude control thrusters, both Juno and Cassini boast relatively high thrust propulsion for conventional large ΔV changes like orbit corrections [25][26]. While they have yet to be flown on a deep space probe ion thrusters are a burgeoning field that may revolutionize deep space flight. As a result, it is appropriate to touch on a few outer space missions that have used ion thrusters for propulsion.

The first mission to utilize an ion thruster was SERT-1 in 1964 [27]. It validated the concept of using ion propulsion in space. This led to the development and launch of Deep Space-1. Deep Space-1 flew the NASA NSTAR thruster, demonstrated the viability of using an ion thruster for interplanetary missions, and paved the way for the Dawn mission [28]. The Dawn mission objective is to explore Vesta and Ceres. This mission required an innovative propulsion system to supply the required ΔV. Dawn used ion propulsion along with a gravity assist to reach the orbits of Vesta and Ceres. Figure 2.9 displays Dawn's exact orbit and the characteristic spiraling of ion engines. In total Dawn will be thrusting for 1100 days [29].

2.4.2   Communications and Data-handling.   Due to the high speed of advancement in the field of electronics for the past two decades, the review on the communications and data handling systems will be restricted to the most recent spacecraft, New Horizons.

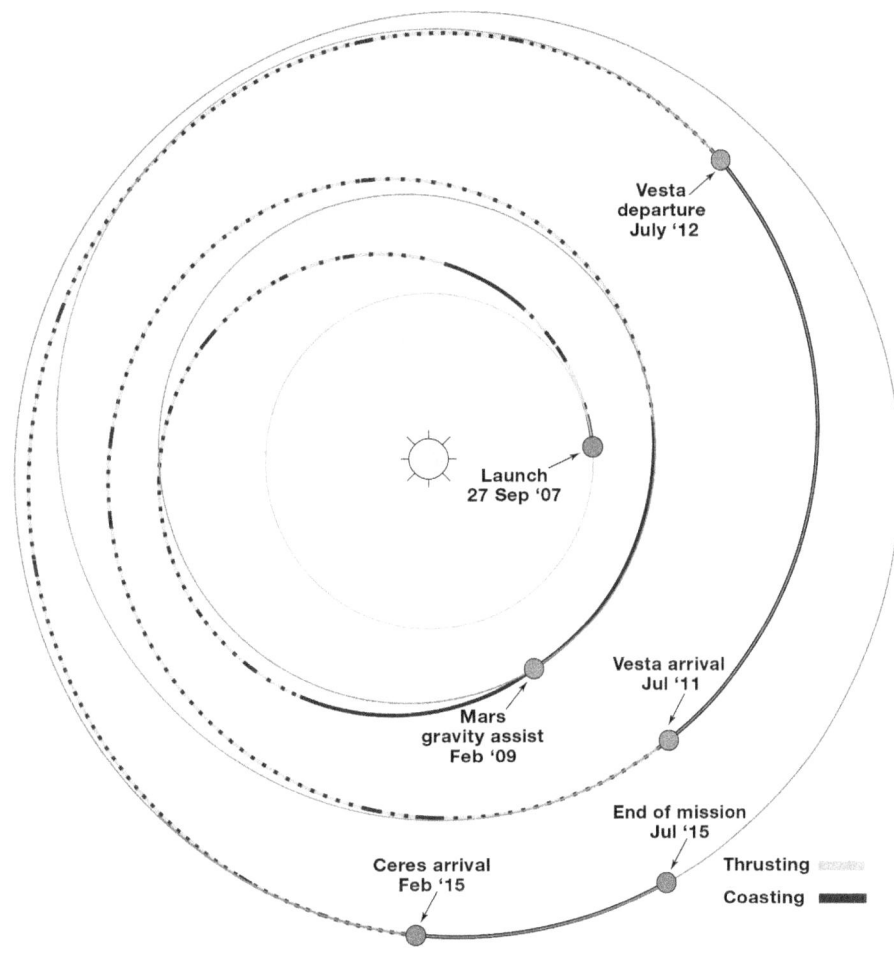

Figure 2.9: Dawn Trajectory [4]

New Horizons is capable of storing 64 Gbits in non-volatile memory [5]. These 64 Gbits are divvied into the 16 independently addressable segments which make up the solid state recorder [5]. The maximum rate that New Horizons can record new data is 13 Mbits/s [5]. After the original recording is complete, data can processed off of the solid state recorder and compressed via either lossless or lossy compression. To further compress the data New Horizons is also capable of subframing whereby the image data can be separated into eight subframes and compressed individually. Paired with New

Horizons' software that can detect which subframe the observation target is on the other subframes can be deleted thus conserving valuable space. Following compression the data can be transmitted back to Earth or rewritten to the solid state recorder. To help organize and control the flow of data New Horizons designates a data type for each recording. There are 51 data types [5]. An example of the data types are compressed science data, instrument data, and spacecraft housekeeping data [5]. Finally, New Horizons can bookmark data. This allows for quick access to commands and data.

The communications subsystem has two primary tasks: command uplink and telemetry downlink. To accomplish these tasks New Horizons employs an antenna assembly, traveling wave tube amplifiers, an ultra-stable oscillator, an uplink command receiver, and a regenerative ranging circuit [5].

The Antenna assembly is composed of a forward and underside antenna system. The forward antenna system possesses a hemispherical coverage low gain antenna (LGA), a high gain antenna (HGA), and a medium gain antenna (MGA) [5]. The LGA is capable of providing close Earth communication. It was able to communicate with Earth until it was 1 AU away [5]. The HGA measures 2.1 meters and was designed to meet the requirement of a 600 bit/s data rate at its mission distance [5]. The mission distance is 36 AU. The HGA provides a 42 dB downlink gain when it is pointed within .3 degrees of the Earth [5]. This means the spacecraft will transmit all 5 Gbits of mission data in 172 days if 8 hour passes are executed every day [5]. This calculation includes a 2 dB margin [5] . Finally, the MGA allows the spacecraft to transmit/acquire Earth at angles up to 4 degrees [5]. The underside antenna system is comprised of a single hemispherical coverage low gain antenna that serves as a redundancy for its twin on the forward antenna system [5].

The ultra-stable oscillator (USO) maintains the spacecraft's time base and provides frequency stability for the uplink and downlink [5]. The USO is measured to have an Allan Deviation of 3×10^{-13} (*unitless*) for a 1 second interval and 2×10^{-13} for an interval of 10 seconds. Allan Deviation is the square root of the Allan Variance where the Allan Variance is "one half the time average of the squares of the differences between successive readings of the frequency deviation sampled over the sampling period" [30]. Thus, a low Allan Deviation is desired because it implies good frequency stability over the period.

The uplink command receiver on New Horizons is innovative. It performs the tasks of command decoding, ranging tone demodulation, and carrier tracking and, unlike its predecessors utilizes a very low-power digital design [5]. Previously uplink command receivers used approximately 12 W of power, but the uplink command receiver on New Horizons consumes only 4 W of power [5]. The 8 Watts saved here is of great importance for a deep space mission and accounts for almost 5% of the entire power budget [5].

Deep space probes are tracked using phase modulated tones sent from the deep space network [5]. The engineers of New Horizons improved this method of tracking by integrating a regenerative ranging circuit into the spacecraft. The regenerative ranging circuit works by using a delay-locked loop to replicate the uplink signal and adjust the spacecraft timing accordingly. This eliminates wide band uplink noise, a major contributor to tracking error, from the signal the spacecraft emits. This means New Horizons can be tracked to within 10 meters until it passes 50 AU [5].

The last pieces of the communications system are the travelling wave tube amplifiers (TWTA). The TWTAs amplify the downlink signal to provide an improved data rate. There are two TWTAs one connected to the MGA and one connected to the HGA [5]. Due the the presence of a hybrid coupler both TWTAs can be connected to the HGA [5]. In this configuration one TWTA will transmit right hand circular and the other will transmit left hand circular [5]. The deep space network architecture has the capability to combine the two signals on the ground. Furthermore, operating the TWTAs in this configuration will increase the date rate by almost 2 times as shown by Figure 2.10 [5].

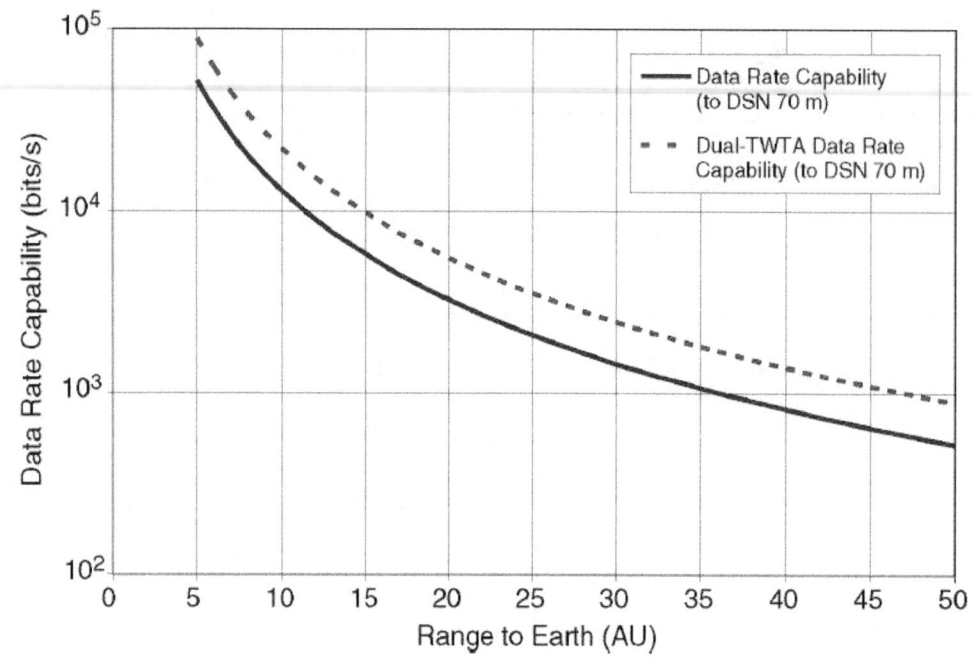

Figure 2.10: New Horizons Data Rate   [5]

An important final note, every deep space probe currently communicates with Earth through the Deep Space Network (DSN). The operating frequencies of the Deep Space Network are shown in Table 2.3.

Table 2.3: DSN Allocated Frequency Band (MHz) [14]

Band Designation	Deep Space Bands (for spacecraft greater than 2 million km from Earth)		Near Earth Bands (for spacecraft less than 2 million km from Earth)	
	Uplink (Earth to space)	Downlink (space to Earth)	Uplink (Earth to space)	Downlink (space to Earth)
S-band	2110–2120	2290–2300	2025–2110	2200–2290
X-band	7145–7190	8400–8450	7190–7235	8450–8500
K-band	*	*	*	25500–27000
Ka-band	34200–34700	31800–32300	*	*

* No allocation or not supported by the DSN.

2.4.3   Electrical Power.   Travelling to deep space puts serious restrictions on power because it makes the use of solar panels infeasible. Even a theoretical, 100 percent efficient solar panel array would output a very small amount of power due to the rapidly declining solar irradiation. Figure 2.11 clearly demonstrates this phenomenon.

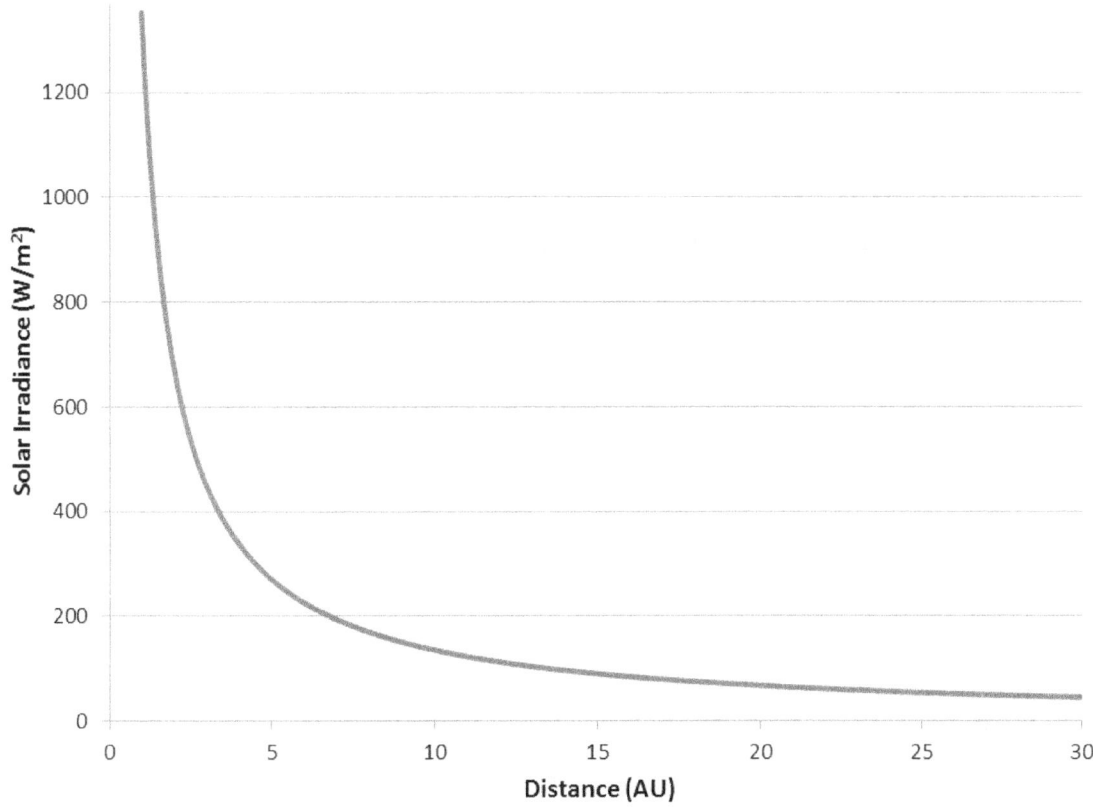

Figure 2.11: Solar Irradiance

Mathematically, the relationship between the solar irradiance and distance is the solar irradiance is inversely proportional to the square of the distance from the sun. At 1 AU the solar irradiance is 1358 W/m^2 however once a spacecraft reaches Jupiter, at approximately 5 AU, the solar irradiance dwindles to a paltry 270 W/m^2 and again it drops precipitously to 135 W/m^2 at 10 AU. As a result, historically, deep space missions have not used solar panels as their primary source of power.

Nonetheless, Juno, NASA's newest deep space satellite, uses solar panels. Figure 2.12 shows the gargantuan size of Juno's solar arrays. Since Juno's mission only involves traveling to and orbiting Jupiter it is possible to use efficient, sizable solar arrays.

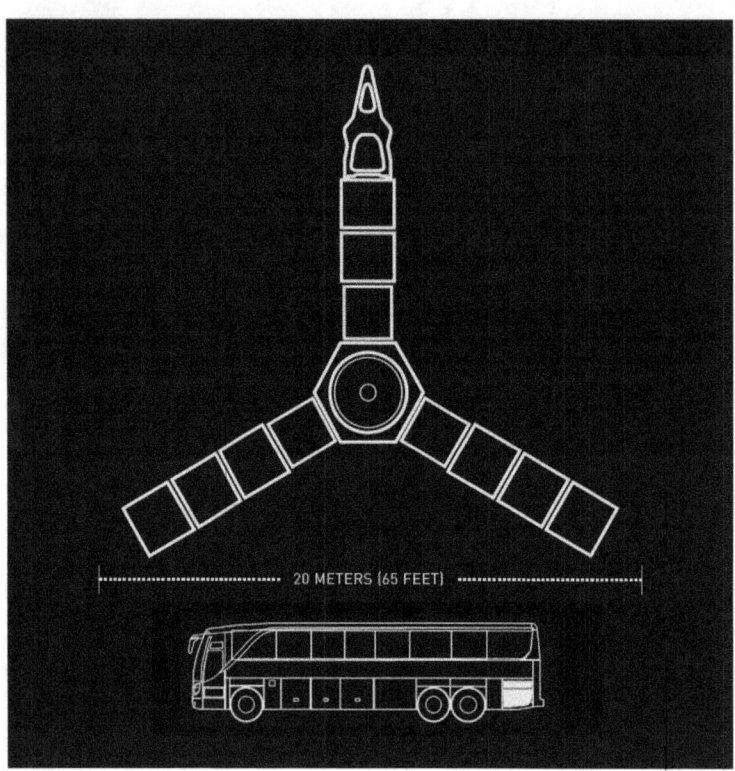

Figure 2.12: Juno Solar Panels   [6]

More specifically, each of Juno's solar panel arrays measure 23.85 m^2 making the total array size 71.55 m^2 [6]. The panels have an efficiency of 28.3 percent [31]. This provides a total power output of approximately 14 kW at Earth and 450 W at Jupiter [6]. However, as the mission progresses the solar panel efficiency will degrade at a rate dependent upon the harshness of the environment. An electron fluence of 5×10^{14} at 1 MeV will result in an eight percent loss of beginning of life power [32]. Ultraviolet rays and impacts (e.g. micrometeors) can cause a 1.7 % and 1% loss in beginning of life power respectively over a twenty year span[32].

While solar panel arrays have been utilized, most of other deeps space missions have opted to use radioisotope thermoelectric generators , or RTGs due to the inefficiency of solar panels in deep space. Radioisotope thermoelectric generators work by converting the heat produced by radioisotopes into electric power. Table 2.4 shows the progression of the RTG.

Table 2.4: RTG Evolution

RTG	Mission (Yr)	BOL Power Output (W)	Power Density (W/kg)	RTG Mass (Kg)	Dimensions
SNAP-3B	Transit 4A (1961) Transit 4B (1961)	2.7	1.28	2.1	Diameter - 12.1 cm Height - 14 cm
SNAP-9A	Transit 5BN-1 (1963) Transit 5BN-2 (1963)	26.8	2.18	12.3	Diameter – 50.8 cm Height – 26.7 cm
SNAP-19B	Nimbus-3 (1969)	28.2	2.10	13.4	Diameter – 53.8 cm Height – 26.7 cm
SNAP-27	Apollo 12 (1969) Apollo 14 (1971) Apollo 15 (1971) Apollo 16 (1972) Apollo 17 (1972)	63.5	3.24	19.63	Diameter – 40 cm Height – 46 cm
SNAP-19	Pioneer X (1972) Pioneer XI (1973) Viking (1975)	40.3	2.96	13.6	Diameter – 50.8 cm Height – 28.2 cm
MHW-RTG	Voyager (1977) LES 8 (1976) LES 9 (1976)	158	4.19	37.69	Diameter – 39.73 cm Height – 58.31 cm
GPHS-RTG	Cassini (1997) New Horizons (2006) Galileo (1989) Ulysses (1990)	245.7	4.25	57.8	Diameter – 42.2 cm Height – 114 cm
MMRTG	Mars Science Lab (2011)	120	2.79	43	Diameter – 64 cm Height – 66 cm
ASRG	N/A	143	7	20.2	Length – 72.5 cm Width – 29.3 cm Height – 41.0 cm

Every one of the RTGs listed in the table used or uses plutonium-238 as its radioisotope. Plutonium-238 has a half-life of 87.7 years and an average power density of 570 W/kg [33]. Figure 2.13 displays the exponential decay of plutonium with time. Therefore, the initial radioisotope mass must be chosen so that it provides sufficient power at the end of life (EOL).

Table 2.4 also shows that great strides have been made in improving the power density over the last fifty years. The Advanced Stirling Radioisotope Generator (ASRG)

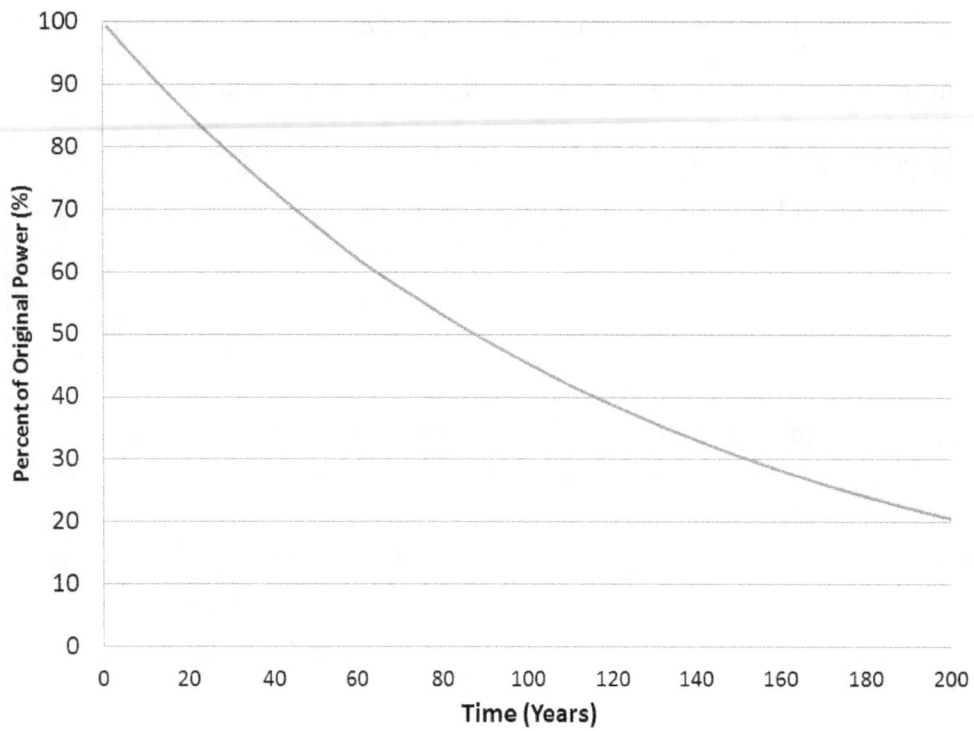

Figure 2.13: Theoretical Decay of Plutonium-238

constitutes a 65% increase in power density from it predecessor the General Purpose Heat Source (GPHS). Interestingly, the Multi-Mission RTG (MMRTG) deviates from the trend of increasing power density. The goal of the MMRTG, unlike the ASRG, was not to maximize power density but to minimize the amount of plutonium-238 used while maximizing the output power. In other words it was engineered to maximize the conversion efficiency. Therefore, it does not follow the general trend.

Over the years deep space electrical power sources have evolved from rudimentary RTGs to highly efficient solar panels at close distances and high power density RTGs in the deepest reaches of space.

2.4.4 Environmental Control. The harsh environment of space (e.g. charged particles, the vacuum of space, radiation) requires the spacecraft to be hardened against a multitude of environmental dangers. Yet, out of all of the subsystems the environmental control subsystem has shown the least evolution over time. The techniques that were utilized on the Pioneer missions are still the primary methods of environmental control. This is due, in part, to the fact that what was optimal in the 1960s is still optimal today. As a result, the vacuum of space presents many unique challenges. The most notable among them is the challenge of heat transfer.

In space, heat can be transferred away from the spacecraft only by radiation. Which is expressed as,

$$q = \epsilon \sigma A(T_{s/c}^4 - T_\infty^4) \quad [34] \tag{2.17}$$

where ϵ is the emissivity, σ is the Stefan-Boltzmann constant, A is the area of the body, $T_{s/c}$ is the temperature of the spacecraft and T_∞ is the temperature of the surroundings. However, heat can be transferred throughout the spacecraft through all three modes–radiation, conduction, and convection. So, the methods of ensuring that the spacecraft does not overheat or freeze are limited.

The most difficult challenge in temperature control is preparing for large temperature swings. The temperature instrumentation on the Ulysses will be examined because the spacecraft travels out to 5 AU and close to the sun so it was engineered to radiate heat as well as be insulated depending on the varying heat flux. Ulysses' instrumentation required that the temperature remain between 5 and 25 degrees Celsius to guarantee that its fuel source did not freeze and its electronics did not overheat [24]. To accomplish this its engineers used standard techniques.

The first technique, insulation, relied on multi-layer insulation blankets, or MLI blankets, which are shown in Figure 2.14 as the golden colored covering. These blankets are layered to help insulate and maintain thermal balance. With that, their outer layer is reflective to decrease the emissivity and absorptivity of the spacecraft. Mathematically,

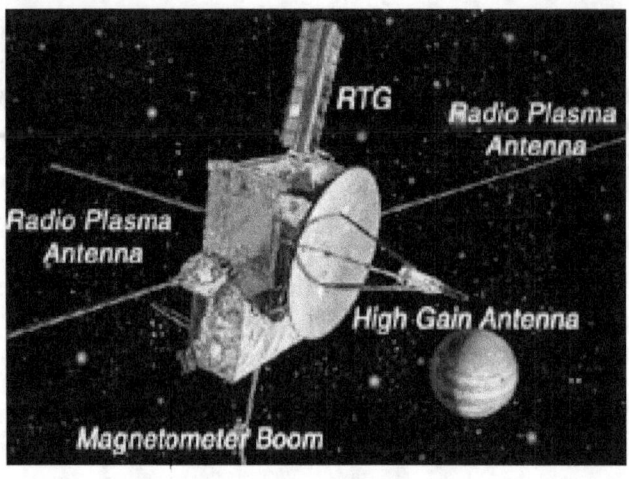

Figure 2.14: Ulysses Spacecraft   [7]

the easiest way to understand their usefulness is to examine the one-dimensional composite heat transfer equations.

$$q = UA\Delta T \quad [34]$$ (2.18)

The blankets use the principles of conduction to limit the flow of heat and maintain an internal balance in the spacecraft. Equation 2.19 shows how the overall heat transfer coefficient U effects the heat transfer rate. In Equation 2.21, A is the surface area and ΔT is the temperature difference between the outside and inside of the composite wall. The overall heat transfer coefficient is dictated by

$$U = \frac{1}{R_{tot}A} \quad [34]$$ (2.19)

and

$$R_{tot} = \Sigma R_t \quad [34]$$ (2.20)

where R_t is the thermal resistance of a single layer of the blanket. The thermal resistance equation for radiation is

$$R_t = \frac{T_{s/c} - T_\infty}{\epsilon A \sigma (T_{s/c}^4 - T_\infty^4)} \quad [34]$$ (2.21)

Consequently, there is direct relationship between the number of layers, their thermal resistances, thermal emissivities, and the heat transfer rate. An increased number of layers, decreased emissivities and higher thermal resistances correspond to a lower heat transfer rate; thus, keeping the satellite warm or cool. Ulysses' MLI blankets were composed of 20 layers of aluminized mylar that possess emissivities in the range of 0.06 to 0.08 [24]. On top of the MLI blankets, Ulysses also possesses a commandable thermal radiator which radiates heat away from the spacecraft.

The second technique for thermal control is optimizing the layout of the spacecraft. Distance provides a barrier to overheating. Using the same principles as the blankets, the more distance between electronic devices the more surfaces the heat must be conducted through. By spacing the electronics in this manner, the entire spacecraft can maintain a steady, uniform temperature.

Finally, though Ulysses did not possess them, radioisotope heater units (RHU) have been used in the past to keep spacecraft warm in the deepest reaches of space. Similar to the aforementioned RTGs, RHUs utilize radioisotopes, but they radiate the heat produced by the radioisotopes and do not convert the heat to electric power. As a result, they are much smaller and very efficient as a long term heat source. Radioisotope heater units also use plutonium-238 [33]. They weigh only 39.7 grams and measure 2.54 cm in diameter and 3.3 cm in height [33]. They output 1 W of heat and when these units are placed properly they can adequately heat a small area [33].

Charged particles present another unique challenge. As spacecraft travel on their journey they are bombarded with charged particles which causes the phenomenon known as spacecraft charging. *Understanding Space* defines spacecraft charging as a build up of charge on different parts of a spacecraft [35]. What is more, this build up of charge is dangerous because it may eventually result in a discharge, which is the natural result when there is a concentration of charge in one place and an absence in another. These discharges can damage the electronics, solar panes, and surface coatings [35]. To combat this, spacecraft are covered with highly conductive materials like Indium Tin Oxide [24]. This conductive coating provides a passive defense against charging by distributing any charge build up so that a discharge will not occur.

Charged particle collisions also result in sputtering. To borrow the analogy from Sellers' text sputtering is like "sand blasting" the spacecraft [35]. A multitude of small collisions between charged particles and the spacecraft cause the spacecraft's exterior to ablate. The materials ablated from the surface often return due to lighter pressure from radiative heat transfer to another–usually cold–part of the spacecraft and cause damage. This in turn results in lower performance from the spacecraft's external sensors and a slow degradation in efficiency of the spacecraft's solar panels [35]. To overcome this, astronautical engineers utilize specific space qualified materials when manufacturing parts. Although the controlled manufacturing processes do not eliminate sputtering, they do minimize it. These specialized parts are also selected to prevent out-gassing and cold welding.

Out-gassing is the expulsion of gasses from material. These gasses, e.g. H_2O, are trapped in the material on Earth because of the pressure exerted by the atmosphere [35]. However, once these materials are introduced into the vacuum environment of space the gasses are expelled. This expulsion can cause arcing and damage the electronics and sensor payload, and is again combated by using space qualified parts [35].

Cold welding is when two parts on the spacecraft fuse together due to having little separation between them. On Earth there may be a free flow of air molecules between the two parts to prevent this phenomenon but once in space the air flow stops and the parts are welded together [35]. Cold welding is combated by using material combinations. Two different materials when placed close together are less likely to fuse together than two similar materials.

The final major challenge of the space environment is radiation. In deep space, the primary source of radiation will come from galactic cosmic radiation. Galactic cosmic radiation (GCR) is made up of high energy nuclei in the MeV to GeV range [36]. The maximum yearly exposure to GCR for a spacecraft is approximately 1.3×10^8 $protons/cm^2$ which is equivalent to 10 $rads/yr$ [36]. The high kinetic energy of the GCR can cause errors and permanent damage to solid state electronics [36]. To guard the electronics from the GCR electronics are space hardened or put in an electronics vault like on the Juno mission, which is made of tantalum [37]. The electronics vault on the Juno mission, which expects a radiation dose greater than 100 Mrad, decreases the exposure the electronics see by almost four orders of magnitude to 25 Krad [37].

2.5   Conclusion

This chapter presented what has been done thus far in satellite systems and introduced the physics of the gravitational lens. By understanding what has been done in satellite systems an approach was developed for the design of the FSM spacecraft, which is to follow in Chapter 3.

3 Analysis

3.1 Introduction

After extensive study on the concept of gravitational lensing between 1987 and 1993, Dr. Maccone, the originator of the idea, submitted a medium-class mission (M3) proposal to the European Space Agency for the design of the Focal Space Mission. This proposal addressed a number of things ranging from the impetus for the mission to the payload requirements. Following his proposal, Dr. Maccone continued to expand his ideas, and in Deep Space Flight and Communications he compiled the most recent information on the topic. In the book, he touches on the physics of gravitational lensing, its enabling capabilities, and the myriad of ways to exit the solar system. However, what is not discussed in real terms are the communication challenges, the specific subsystems requirements, the subsystems themselves, an in-depth ΔV analysis, trajectory analysis, and possible launch vehicles. Dr. Maccone presents a very thorough scientific analysis but does not delve into the engineering required and that is the heart of the next two chapters.

3.2 Launch Vehicle

The launch vehicle analysis for this mission comes down to one thing–C_3. C_3 is the term used to describe the excess energy provided by the launch vehicle. It is defined by the equation

$$C_3 = V_\infty^2 \quad [15]$$

(3.1)

where V_∞ is the hyperbolic excess velocity. The hyperbolic excess velocity is the velocity difference between the escape velocity and the velocity of the spacecraft. The Earth escape velocity for a circular orbit is given by

$$v_{esc} = \sqrt{\frac{2GM_E}{r}} \quad [15]$$

(3.2)

where G is the gravitational constant, M_E is the mass of the Earth, and r is the orbit radius.

33

When the same launch vehicle configuration is considered a higher C_3 implies a higher initial mission velocity but a lower delivered mass. The remainder of this section will explore the optimization of ΔV as a function of C_3. In more specific terms, the goal is to find the launch C_3 that maximizes the mission ΔV.

The first step in this examination was to look at the ideal rocket equation

$$\Delta V = I_{sp} g_0 \ln\left(\frac{m_0}{m_f}\right) \quad [38] \tag{3.3}$$

where I_{sp} represents the specific impulse, g_0 is the acceleration due to gravity at the Earth's surface, m_0 is the initial mass, and m_f is the final spacecraft mass. The initial mass is equal to the sum of the propellant mass m_p and the final spacecraft mass. Thus clearly an increase in propellant mass with no change to the final mass will result in a logarithmic increase in ΔV. Also, an increase in specific impulse will result in a linear increase in ΔV. With this knowledge it is now appropriate to look at trends in the excess energy.

Quickly looking at Figure 3.1 the trend is unmistakable. As mentioned earlier, for a given launch vehicle configuration the larger the payload mass the lower the excess energy. Therefore, if the final spacecraft mass is assumed to be the minimum mass required to accomplish the mission then the only mass that is varying is that of the propellant. So an increase in propellant mass corresponds to a decrease in excess energy.

Exploring the concept in more depth, the launch vehicle configuration, similar to the spacecraft, operates under the constraints of the ideal rocket equation. This means the ΔV of the launch vehicle is also a function of the specific impulses, final masses and propellant masses of the launch vehicle configuration. Therefore, the optimization of the ΔV as a function of the C_3 is dependent only upon these values for the launch vehicle and the spacecraft. With that, those values are constants for a given launch vehicle configuration and are what produce the curves in Figure 3.1. As a result, the optimization is constrained to the curve meaning it is now dependent solely on the spacecraft's values

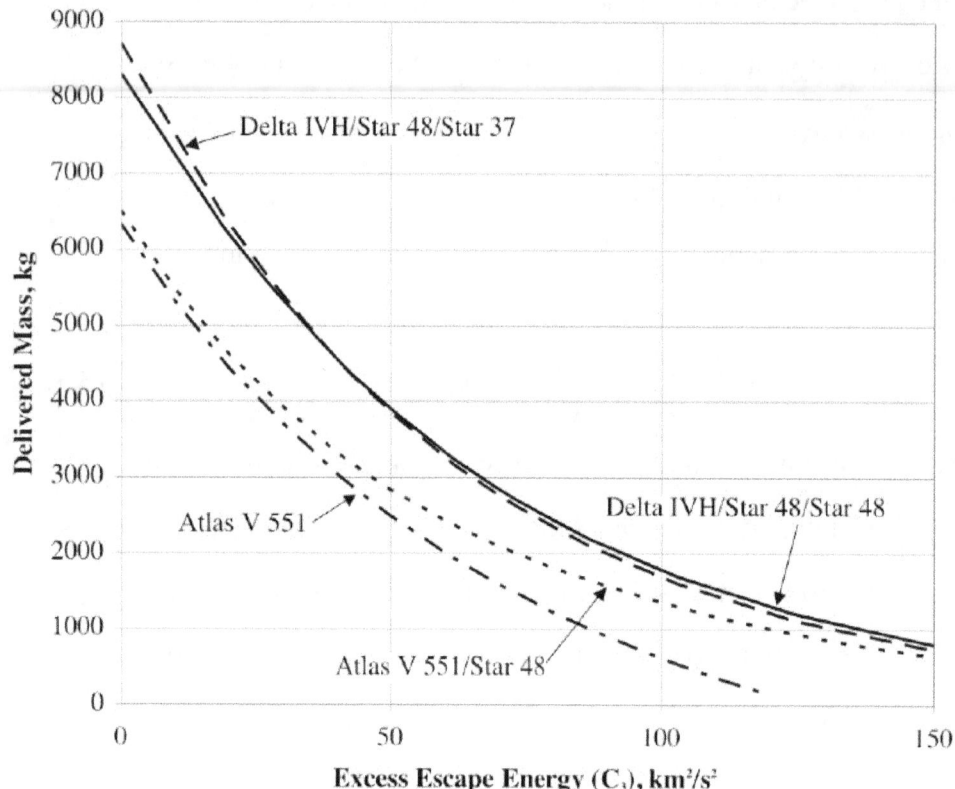

Figure 3.1: Selected Launch Vehicle Performance   [8]

for specific impulse, propellant mass, and final mass. Interestingly, the spacecraft's propellant mass and final mass are also constrained by the launch vehicle curves because the curves specify the delivered mass, which is the sum of the propellant mass and final mass. This leaves the specific impulse of the spacecraft as the only free variable.

Figure 3.2 shows the scaling effect that the specific impulse has on the ΔV. An increased specific impulse results in a linear increase in the ΔV for a given mass ratio which is defined as

$$M_r = \frac{m_0}{m_f} \tag{3.4}$$

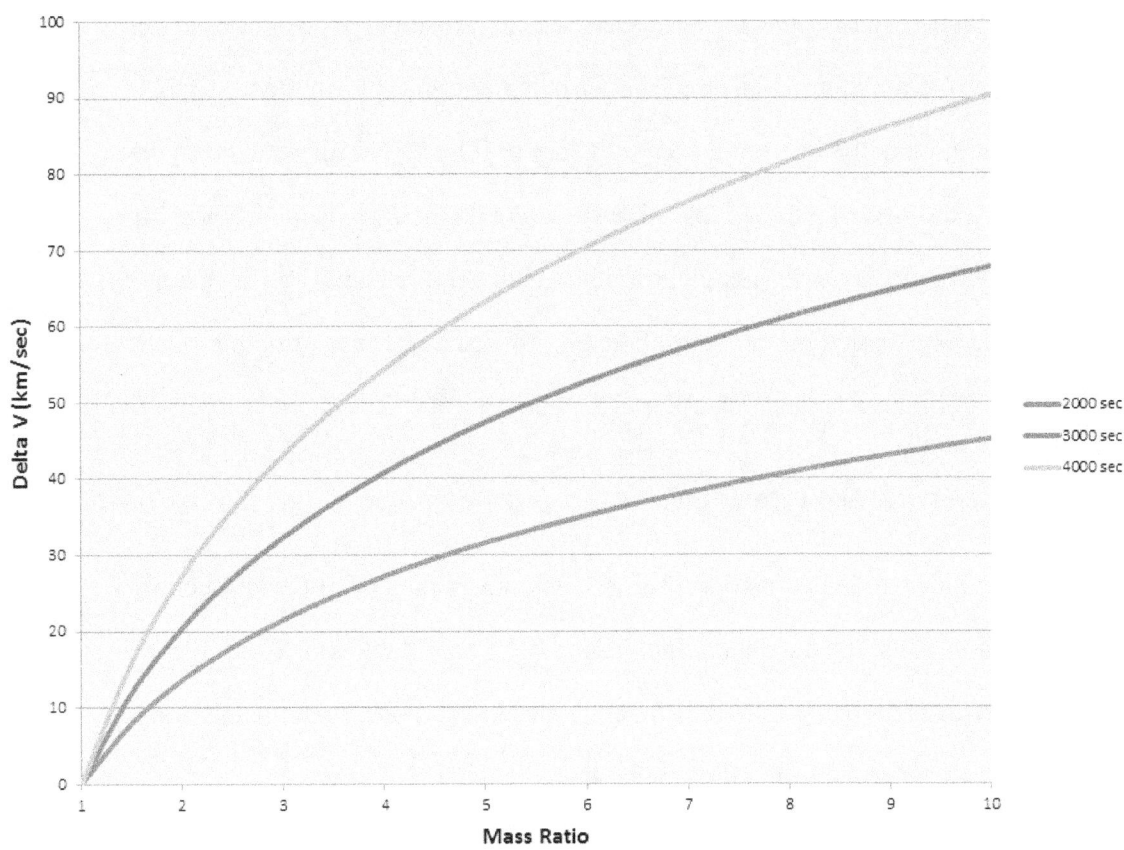

Figure 3.2: Effect of Specific Impulse on ΔV

Due to this fact, if the spacecraft has a higher specific impulse than the launch vehicle upper stage then optimizing the ΔV by varying the C_3 requires the C_3 to be minimized because the spacecraft will use the mass more efficiently. Upper stages have low specific impulses in the hundreds, e.g the Star 48 has a specific impulse of 289.9 seconds, compared to electric propulsion options with thousands of seconds of specific impulse [39] . Hence, the presence of an electric thruster, e.g. the NSTAR thruster mentioned in the literature review, in the range of thousands of seconds would imply that to optimize launch ΔV , C_3 should be minimized. However, if a thruster is not present or it will only be used in orbit corrections/gravity assists the opposite is true. The optimum C_3 is the maximum C_3.

36

The optimum launch vehicle for the first case (electric thrusters are present) will be the one that provides the maximum delivered mass to the minimum C_3. In regard to this mission, the minimum C_3 is the parabolic excess energy of 0 km^2/s^2. This is taken as the minimum to avoid spiraling around the Earth. For the second case (thrusters only used for orbit corrections and/or gravity assist), the optimum launch vehicle would be the one that delivers the spacecraft mission mass to the highest C_3. So based on these principles after a propulsion system is chosen a launch vehicle may be chosen.

3.3 Communications and Data Handling

The communications and data handling subsystem gathers data from the spacecraft, processes the data and transmits the data to the ground. The communications and data handling subsystem also receives commands from the ground station. The transmission and receipt of data is analyzed through the link budget.

The link budget is the relationship between data rate, transmission power, transmission path and antenna size. It is best illustrated by the energy per bit to noise power spectral density ratio, which is given in the following equation where P is the transmitter power, L_l is the transmitter-to-antenna line loss, G_t is the transmit antenna gain, L_s is the space loss, L_a is the transmission path loss, G_r is the receiver antenna gain, k is Boltzmann's constant, T_s is the system noise temperature, and R is the data rate in bits per second.

$$\frac{E_b}{N_o} = \frac{PL_lG_tL_sL_aG_r}{kT_sR} \quad [15] \tag{3.5}$$

The transmitter power is an input parameter with units of W. The amount of transmitter power available is dependent upon the choice of the power system. However, the power system has yet to be determined at this point in the analysis. As a result, a range of 1 to 300 W was examined throughout this section.

The transmitter-to-antenna line loss represents the communication losses that occur when data is transferred from transmitter on the spacecraft to the spacecraft antenna. On average, this loss is .8, a unitless value [15].

The transmitter and receiver antenna gains are given by the equation

$$G = \frac{\pi^2 D^2 \eta f^2}{c^2} \ [15] \tag{3.6}$$

where π is the familiar mathematical constant, D is the diameter of the antenna, η is the antenna efficiency, f is the frequency, and c is the speed of light. Thus an increase in antenna size will lead to a parabolic increase in gain, as will an increase in frequency. The antenna efficiency is a function of the antenna shape. Three of the most common antenna types are the parabolic reflector, helix, and horn which have efficiencies of .55, .7, and .52 respectively. Finally, the frequency of the transmission is dependent upon the channels available in the DSN. Each channel has a bandwidth of approximately 370 kHz and falls within the frequency ranges shown in Table 2.3. Though the size of the spacecraft antennas have yet to be determined in this analysis the DSN receiver antennas are parabolic and 70 meters in diameter [14].

For deep space missions the space loss has a gargantuan effect on the signal to noise ratio. Given by the equation

$$L_s = \left(\frac{c}{4\pi S f}\right)^2 \ [35] \tag{3.7}$$

the space loss represents the attenuation of the transmitted signal due to travelling through free space. In Equation 3.7, c is again the speed of light in meters per second, π is the familiar constant, S is the distance from the antenna to the ground receiver and f is the frequency of the signal. Therefore, the further the spacecraft travels from Earth the greater the loss as shown in Figure 3.3 . Notice that Figure 3.3 is plotted on a logarithmic scale in the vertical axis. Finally, for the figure, the frequency was held constant at 2291.66 MHz, one of the DSN frequencies. Furthermore, this will be the assumed FSM frequency.

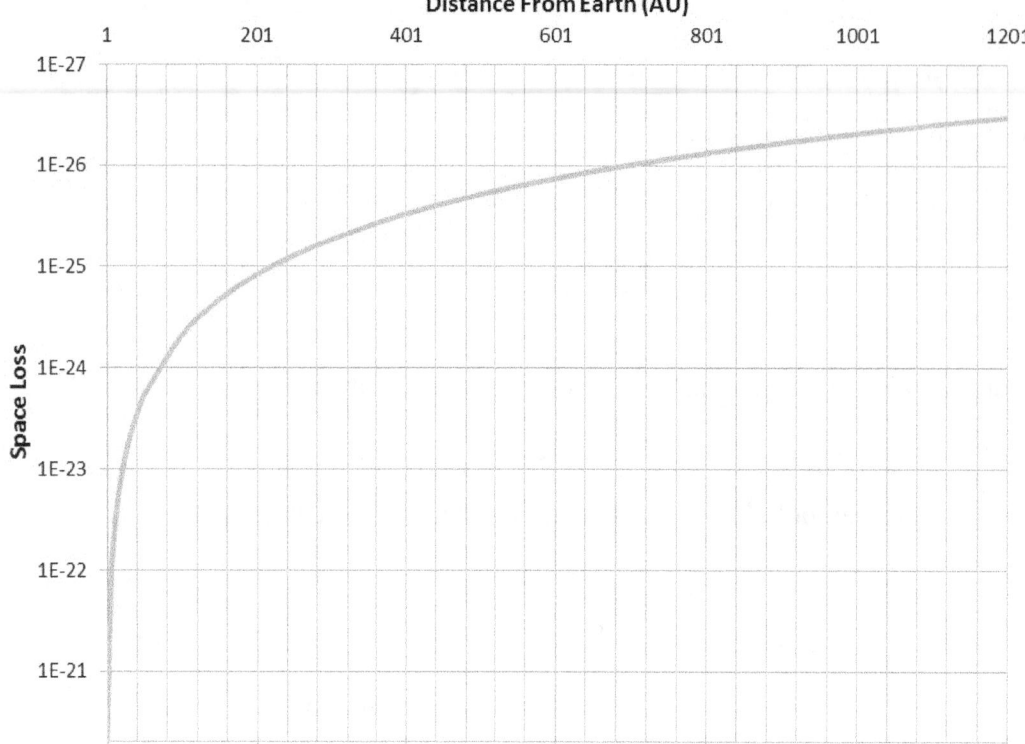

Figure 3.3: Space Loss

The transmission path loss is variable, and it is based on the conditions of the signal medium. *Space Mission Analysis and Design* by Larson and Wertz outlines the sources of transmission path loss: man-made noise, solar noise, galactic noise, atmospheric noise, quiet sun, sky noise, and black body radiation [15].The sources that have the greatest affect on the assumed frequency of 2291.66 MHz are black body, galactic noise, and quiet sun. The total effect of all sources results in a transmission path loss of ten percent meaning L_a is equal to .9.

The data rate, which has units of bits per second, is the rate information is transferred over a communications link. This rate is determined based on the modulation which will be discussed later in this section.

The final piece of the E_b/N_o ratio is the noise spectral density. The noise spectral density is the Boltzmann Constant, $1.38 \times 10^{-23} \frac{m^2 kg}{s^2 K}$, multiplied by the system noise temperature. The system noise temperature is dependent upon the frequency. For the DSN ranges the uplink and downlink system noise temperatures are 614 K and 135 K respectively.

3.3.1   Shannon Limit.   In the E_b/N_o ratio the data rate is limited by the Shannon-Hartley Theorem. The theorem specifies that there exists a maximum rate of information transfer for a given bandwidth, which is dictated by the expression

$$R_{max} = B \log_2 \left(1 + \frac{(EIRP)L_s L_a G_r}{kT_s B} \right) \ [15] \tag{3.8}$$

where the *EIRP* is the effective isotropic radiated power in watts; L_s, L_a, k, T_s, and G_r are the same as in Equation 3.5; and *B* is the bandwidth. The Shannon Limit is being explored because, with the enormous space loss term at the mission distance, it is important to ensure that the maximum data rate is large enough that data can be transferred in a realistic amount of time. If it is not, then the mission is not viable.

To calculate the maximum data rate an *EIRP* of 10000 *W* was assumed.This corresponds to an approximate antenna diameter of 3 *m*, transmitter power of 4 *W*, frequency of 2291.66 MHz, transmitter efficiency of .55 and line loss of .8. The equation for the *EIRP* is

$$EIRP = PL_l G_t \ [15] \tag{3.9}$$

where the definition of the variables have been previously provided. Also, since the relevant case is the worst case the mission distance was taken to be 1000 AU, the EOL distance. The Shannon Limit data rate at 1000 AU given the values stipulated earlier in this section, a bandwidth of 100 kHz and using Equation 3.8 is 44.65 *kB/sec*. This rate is sufficient. The *EIRP* and mission distance were varied in Figure 3.4 to produce a more detailed picture of the Shannon Limit . With that, the figure displays the positive

40

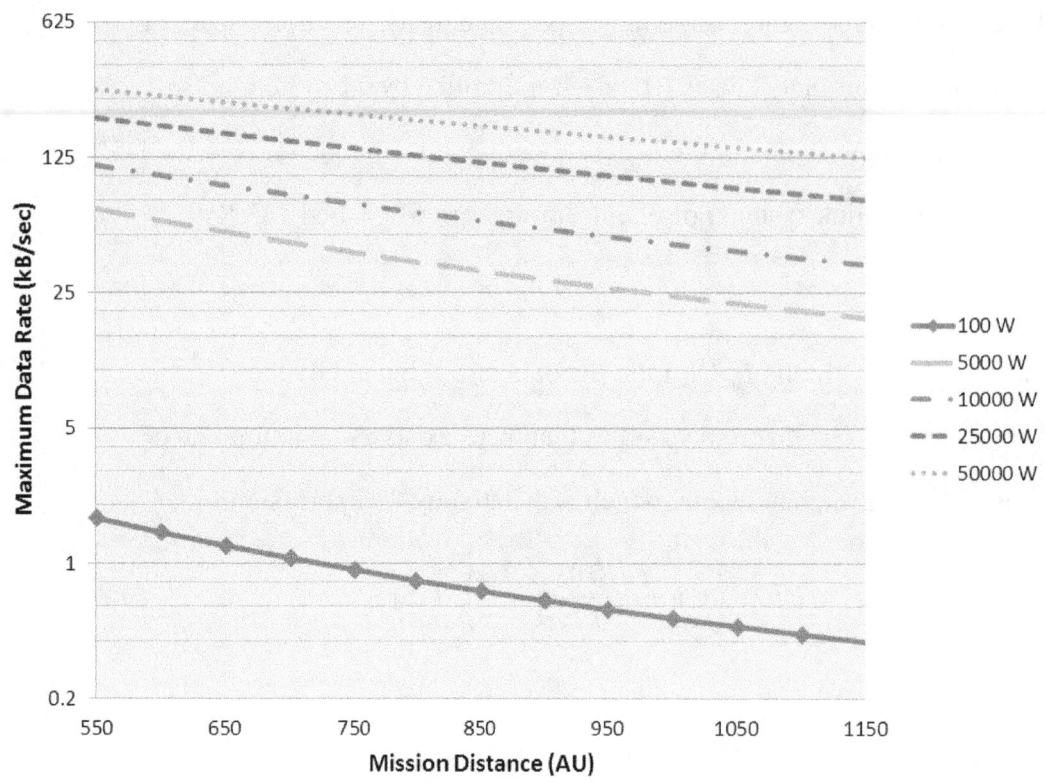

Figure 3.4: Shannon Limit

correlation between *EIRP* and the Shannon Limit as well as the negative correlation between the mission distance and the Shannon Limit. Moreover, Figure 3.5 shows the constant *EIRP* curves. These curves demonstrate how each *EIRP* in Figure 3.4 can be achieved by varying the transmitter power and antenna diameter. While the graph describes configurations that would be undesirable such as requiring 197 W for a 1-meter antenna to achieve an *EIRP* of 50000 W, it is also provides reasonable options such 25 W to an 8-meter antenna to produce the same *EIRP*.

Concluding this subsection on the Shannon Limit, it is important to note that the Shannon Limit is a theoretical limit that can be approached but not achieved. It is approached through the use of modulation and coding schemes. The modulation and coding scheme chosen dictates the actual data rate and bandwidth.

41

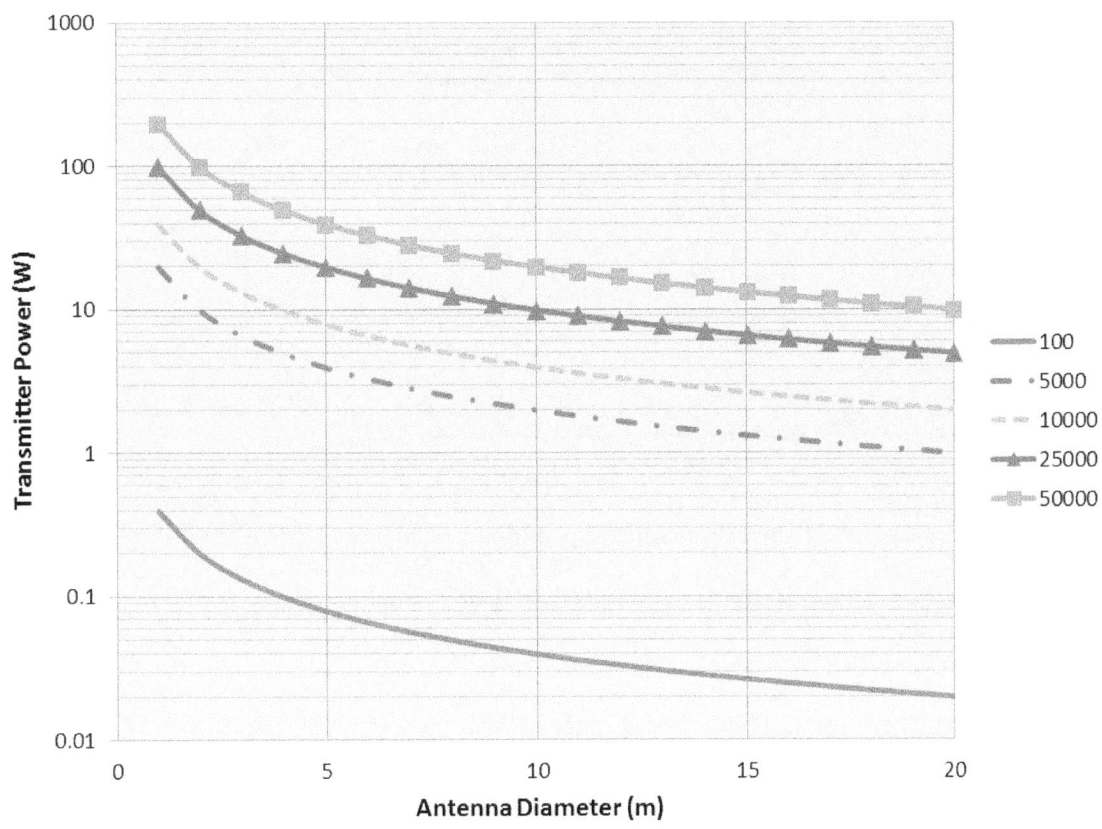

Figure 3.5: Constant EIRP Curves (W)

3.3.2 Signal Modulation. Modulation allows information to be conveyed through radio frequencies. The four types of modulation are polarization, frequency, phase, and amplitude [15]. Spacecraft primarily use frequency and phase modulation due to improved power efficiency [15]. Table 3.1 shows the most common modulation schemes. The second column of Table 3.1 displays the E_b/N_o required for each modulation scheme to achieve a bit error rate (BER) of 10^{-5}. By definition the BER "gives the probability of receiving and erroneous bit" [15]. In regard to Table 3.1, a BER of 10^{-5} implies that the data transmitted will have only one error in every 10^5 bits. This is an acceptable BER.

The best modulation scheme for this mission is the one that provides the best BER performance. The best performance is defined as the one which minimizes the E_b/N_o.

42

Table 3.1: Modulation Schemes

Modulation	E_b/N_o for BER = 10^{-5} (dB)
BPSK	9.6
DPSK	10.3
QPSK	9.6
FSK	13.3
8FSK	9.2
BPSK and QPSK Plus R-1/2 Viterbi Decoding	4.4
BPSK and Plus RS Viterbi Decoding	2.7
8FSK Plus R-1/2 Viterbi Decoding	4.0
MSK	9.6

Though not listed in the table, the Shannon Limit/minimum E_b/N_o for a BER of 10^{-5} is -1.5 dB [15]. This means that the minimum available E_b/N_o of 2.7 dB is only 4.2 dB greater than the Shannon Limit.

The binary phase shift keying (BPSK) and plus Reed-Solomon (RS) Viterbi Decoding technique works by concatenating modulation and coding. First, BPSK modulates the phase of the transmitted signal between two settings—0 and 180 degrees [15]. These two phases represent 0 and 1 respectively [15]. Next, the signal is encoded using forward error correction. Forward error correction is implemented to "reduce the E_b/N_o requirement" [15]. In this case the method of forward error correction is convolutional coding with RS Viterbi decoding. The rate of convolutional coding is 1/2, meaning that for every data bit two bits are transmitted [15]. Finally, when using BPSK modulation the bandwidth is equal in magnitude to the data rate [15].

Rearranging Equation 3.5 produces Equation 3.10, which is an equation for the data rate as a function of *EIRP* and distance since all of the other variables in the equation are constants.

$$R = \frac{(EIRP)L_sL_aG_r}{kT_s\dfrac{E_b}{N_o}} \tag{3.10}$$

The E_b/N_o used in calculating the data rate is a unitless 1.862 which is produced after converting 2.7 dB into its non-dimensional variant. A graph of Equation 3.10 for a range of *EIRP* and distances is shown in Figure 3.6 to give a clearer picture of the relationship.

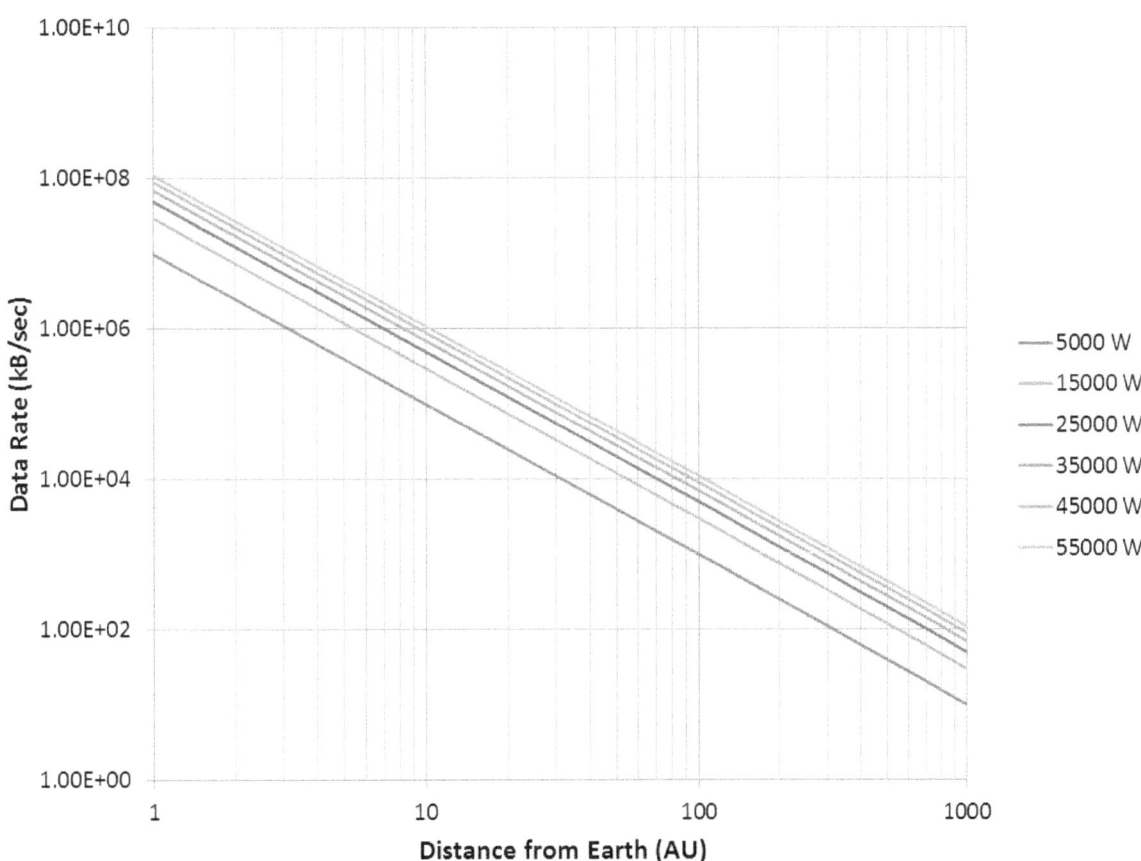

Figure 3.6: Mission Data Rate

In conclusion, based on the size, type and power of the antennas flown the mission data rate can be found using Equation 3.10. The size, type and power of the antennas also affect other communications aspects which will be explored in the following subsection.

3.3.3 Antennas.   In the Background and Theory the communications architecture of previous spacecraft were discussed. Most spacecraft had multiple antennas of different

44

types. Spacecraft antennas fall into three categories high gain antennas (HGA), medium gain antennas (MGA), and low gain antennas (LGA). High gain antennas possess high data rates and narrow half-power beamwidths (HPBW). On the other hand, low gain antennas have wide HPBW but low data rates. Finally, the medium gain antennas make up the space between the extremes. Each antenna type is suited for a different role. The HGA is suited to high bit rate transmissions of mission data. Low gain antennas are primarily used for transmitting data at near Earth distances. On the other hand, medium gain antennas are used to acquire Earth and initiate the link between the spacecraft and the ground station.

Equation 3.6 displayed how the gain is calculated for a parabolic antenna. Hence, to manipulate the gain the engineer must manipulate the diameter of the antenna. However, changing the diameter also affects the HPBW. The HPBW is dictated by the expression

$$HPBW = \frac{21}{f_{GHz}D} \tag{3.11}$$

where f is the signal frequency, D is the antenna diameter and the $HPBW$ is measured in degrees. The HPBW is defined as the "angle across which the gain is 50 percent of the peak gain" [15]. From Equation 3.11 and Equation 3.6 it is clear that the HPBW and gain are inversely related. Therefore, the higher the gain the more narrow the beamwidth and vice versa.

The data rate, HPBW, and gain will be the driving factors in the design of the spacecraft antennas, but the antennas are just one piece of the communications and data handling subsystem.

3.3.4 Autonomy and Lifespan. Radio frequency communications travel at the speed of light so latency becomes a problem with increasing distances. At 550 AU it would take approximately 3.16 days for a signal from the spacecraft to reach the Earth. As a result, the spacecraft will require a high level of autonomy and a lot of storage space.

Autonomy requires a very capable processor . An example processor is the BAE RAD750. Flown on the Juno and Curiosity missions, it is capable of operating at 133 MHz and withstanding 1 Mrad of radiation. What is more, the RAD750 has 490 year mean time between failures [40]. Finally, it has a mass of only 9 g and a power dissipation of 5 W at 133 MHz [40].

To store the flight data the spacecraft will use the same proven set-up as New Horizons, which was outlined in the Background and Theory. New Horizons utilized a 64 Gbit solid state recorder with recording capabilities of 13 Mbits/sec. Unlike New Horizons the FSM will use a radio telescope instead of high resolution optical telescopes; thus, its data requirements will be lower. That was the reasoning behind this course of action. Consequently, the mission data rate requirement will be 500 bits/sec based on the payload data rates. This will allow 5 Gbits of mission data to be downloaded in approximately 240 days with 12 hour passes.

3.4 Power

There are two primary requirements in powering the FSM that will dictate what power systems may be used. The first requirement is to provide sufficient power to operate the mission at the EOL. The second requirement is to provide enough power to propel the spacecraft during the boost phase of the mission.

At the EOL the spacecraft will be at least 800 AU from the sun; thus, solar power will not be an option. That leaves the options of nuclear thermal power and radioisotope electric power. Nuclear thermal can be eliminated from contention immediately because it does not meet the COTS requirement.

The three major RTGs are the MMRTG, GPHS-RTG, and the ASRG. The first two were briefly mentioned in the Background and Theory and have previously been flown. The ASRG is currently in development at NASA Glenn Research Center. Its TRL, or Technology Readiness Level, is TRL 6. A TRL of 6 means that a prototype has been tested to high fidelity in a laboratory. Thus the ASRG is undergoing lifetime and performance testing and does not classify as COTS. Table 2.4 compares the two remaining power options based on mass, dimensions, and power density.

Both of the power sources utilize Pu-238 which has a half-life of 87.7 years. As a result, the mission duration should account for this. Meaning, if possible, the spacecraft should arrive and have transmitted a large amount of data before this time is reached. The degradation in power will also result in lower data rates and limited spacecraft operation.

Finally, the second requirement, which is to provide power for the boost phase of the spacecraft, does not preclude the use of a solar panel array along with the RTG.

3.5 Thruster

3.5.1 Introduction. The thruster analysis is split into four parts. The first part of the analysis examines the spacecraft ΔV and how it can be optimized to produce the minimum trip time . The second piece of the analysis of the payload mass fraction. After that, the analysis addresses thruster efficiency and finally gravity assists.

3.5.2 ΔV. The ΔV, or velocity change required to reach a certain target in a desired amount of time, analysis is an integral part of the spacecraft thruster analysis because it determines the level of performance required by the thruster.

The FSM ΔV analysis is built on the assumption that the satellite is initially placed in a 270 km circular orbit, and that corresponds to an orbital velocity of 7.74 km/s. This assumption is based on the fact that placing a satellite in an orbit of 270 km is well within the capabilities of modern launch vehicles. From there, the satellite is impulsively boosted into an Earth escape trajectory by its upper stage. At 270 km, the minimum velocity to achieve Earth escape is 10.95 km/s, which can be found using Equation 3.2. Thus a ΔV, of 3.21 km/s is required. Furthermore, since these numbers are based off of an assumption it is important to bound its effect. If the assumed altitude was the minimum acceptable low earth orbit altitude of 200 km then the ΔV would be 3.22 km/s and if the insertion orbit was 400 km the ΔV would be 3.17 km/sec. So these values give credence to the assumption by demonstrating that only a small deviation occurs due to the assumed altitude of 270 km.

This new velocity, 10.95 km/s, corresponds to a specific mechanical energy, ϵ, of 119.9 km^2/s^2. Another one of the assumptions of this analysis, and a general assumption in ΔV calculations, was that the specific mechanical energy, ϵ, of an orbit remains constant. As a result, it is easy to calculate the spacecraft's velocity at the termination of Earth's sphere of influence, whose location is determined by Equation 3.12, by plugging r_{SOI} in for r in Equation 3.13 and solving for v.

$$ r_{SOI} = a_{planet} \left(\frac{m_{planet}}{m_{Sun}} \right)^{2/5} \quad [35] \qquad (3.12) $$

$$ \epsilon = \frac{v^2}{2} - \frac{\mu}{r} \quad [15] \qquad (3.13) $$

If, like postulated, the spacecraft left at exactly the escape velocity it would arrive at the sphere of influence with a parabolic trajectory and velocity of 0 km/s relative to the Earth. However, it would still possesses a velocity of 29.78 km/s relative to the sun. Not surprisingly, this is because 29.78 km/s is the orbital velocity of the Earth relative to the sun. Hence, for all intents and purposes the spacecraft is assumed to be in Earth's exact orbit. This is a good simplifying assumption since the Earth's sphere of influence is minuscule relative to 1 AU. More specifically, Earth's SOI is only .006% of 1 AU

As discussed in the launch vehicle section, it is possible for the spacecraft to arrive at the sphere of influence with a velocity higher than 0 km/s relative to the Earth. This velocity is known as the hyperbolic excess velocity, V_∞. This requires a correspondingly higher ΔV. This new ΔV can be found with the following equation.

$$\Delta V = \sqrt{119.05 + V_\infty^2} - \sqrt{\frac{\mu_E}{r}} \qquad (3.14)$$

In equation 3.14, V_∞ represents the desired hyperbolic excess velocity, r represents the initial circular orbit radius and μ_E represents the gravitational parameter of Earth. Figure 3.7 shows the ΔV required to achieve an excess velocity in the range from 0 km/s to 20 km/s for an initial orbit radius of 270 km.

An important excess velocity to look at is 11.99 km/s, or 41.77 km/s relative to the Sun. This is the escape velocity for the Sun at Earth, which can be found by substituting the mass of the Sun for the mass of the Earth and the radius from the the Sun to the Earth in Equation 3.2. The resulting ΔV to achieve solar escape is 8.47 km/s. So, with this ΔV the spacecraft would have enough velocity to escape the sun's gravity; thus, it could coast all the way to the mission distance without a thruster. Finally, with these numbers and equations in hand the author set up a MATLAB script that would allow for the analysis of a ΔV trade space.

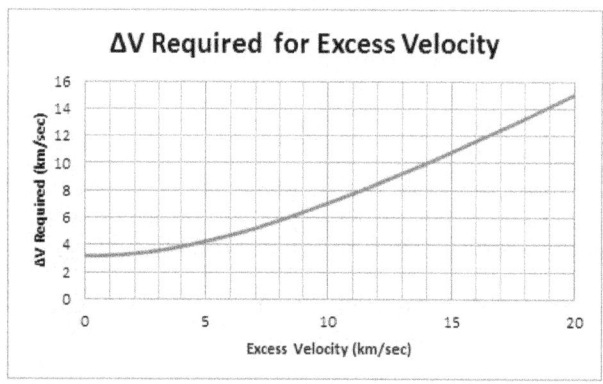

Figure 3.7: Delta-V Required for Excess Velocity

The goal of the aforementioned script was to input the excess velocity, the initial spacecraft mass, the final spacecraft mass, thrust, and the I_{sp}; and output ΔV, payload mass fraction, and time of flight.

For this mission direct impulsive/chemical burns are not a viable option because of the propellent mass requirements inherent in yielding a ΔV in the km/s range. So by necessity, the low thrust/high specific impulse trade space was explored.

Unlike the high thrust/approximately impulsive case where explicit analytical equations are used, the low thrust case directly utilizes the equations of motion.

$$\dot{r} = u \text{ [41]} \tag{3.15}$$

$$\dot{u} = \frac{v^2}{r} - \frac{\mu}{r^2} + A\sin\phi \text{ [41]} \tag{3.16}$$

$$\dot{v} = \frac{uv}{r} + A\cos\phi \text{ [41]} \tag{3.17}$$

Equations 3.15, 3.16, 3.17, are the polar representation of the two dimensional equations of motion. In these equations r expresses the distance from the gravitational body, u is the radial velocity, v is the circumferential velocity, A is the acceleration due to thrust, and ϕ is the polar thrust angle. The ΔV is calculated by integrating the equations of motion over time until the desired radial distance is achieved and using the Ideal Rocket Equation, Equation 3.3 . Using the equations of motion, the Ideal Rocket Equation, MATLAB ode45 (a numerical integrator), and by varying the aforementioned inputs Figure 3.8 and Figure 3.9 were created. In both Figure 3.8 and Figure 3.9 the ΔV was manipulated by changing the propellant mass and keeping the spacecraft mass constant. Finally, the inputs were varied based on attainable and demonstrated thrust levels and specific impulses.

Figure 3.8 displays the effect of varying the thrust level on trip time to 550 AU. For this data set the specific impulse was held at a constant 3000 sec. From the figure it is clear that an increase in thrust will result in a requisite decrease in trip time. Another fact that can be gleaned from this figure is that for every thrust level, at a given specific impulse, there exists an optimal ΔV which minimizes the trip time. This fact is harder to discern but equally valid. Therefore, to optimize trip time one must locate the optimal

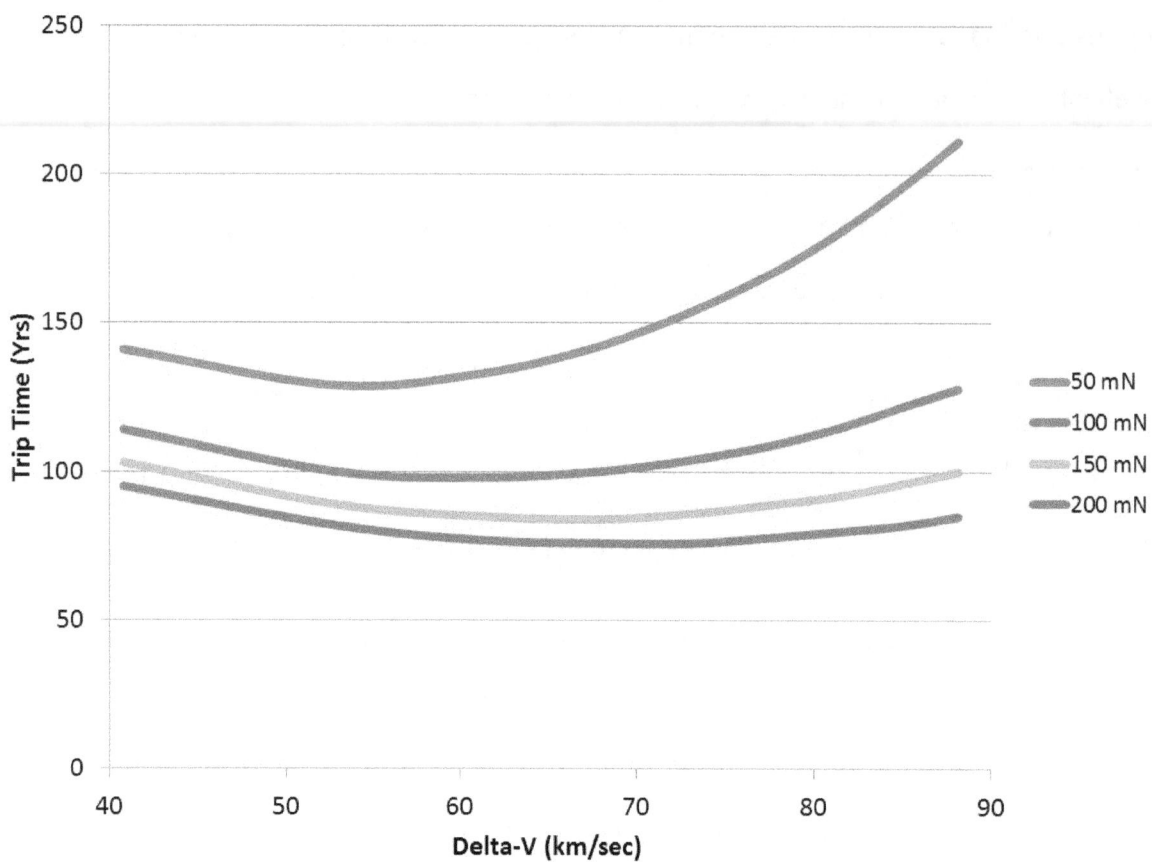

Figure 3.8: Delta-V for Various Thrust Levels

point for a given specific impulse and thrust level, and use that point to find optimal ratio between the final and initial masses.

Figure 3.9 displays the effect of varying the specific impulse on trip time. For this data set the thrust was held at a constant 100 *mN*. It is evident that an increase in specific impulse will result in a decrease in trip time. Also, similar to Figure 3.8, there exist an optimal ΔV. Notice that the optimal ΔV is the same for the 3000 sec series in Figure 3.9 and the 100 *mN* series in Figure 3.8, as they should be.

Conclusively, the MATLAB script along with the data on available thrusters was used to calculate the mission ΔV.

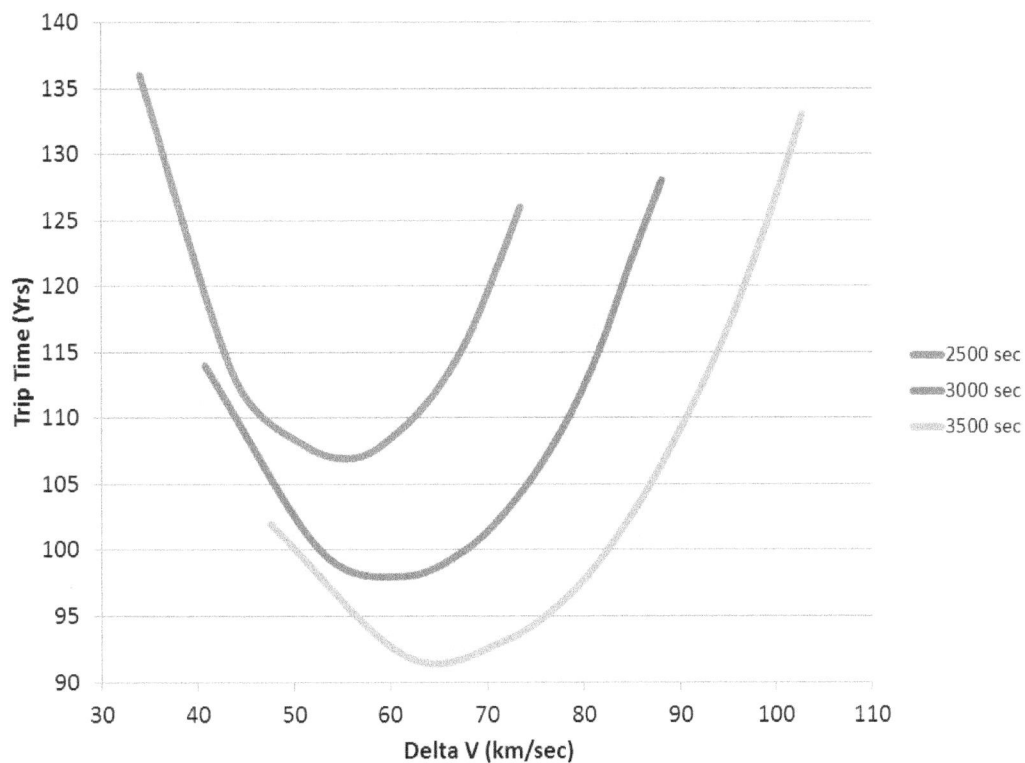

Figure 3.9: Delta-V for Various Specific Impulses

3.5.3 Payload Mass Fraction. The payload mass fraction represents the percentage of the spacecraft that remains for payload after the power source and propellant. The equation for the payload mass fraction is

$$f_{pay} = \frac{m_f - \dfrac{\alpha_s(I_{sp}g_0)^2}{2\eta_t\tau}(m_i - m_f)}{m_i} \quad [15] \tag{3.18}$$

where α_s is the power system specific mass , η_t is the thruster efficiency, m_i is the initial mass , m_f is the final mass , and τ is the burn time. The specific mass has a direct linear relationship with the payload mass fraction. On the other hand, the thruster efficiency has more complicated relationship with the payload mass fraction. An increase in thruster efficiency causes an asymptotic approach to a maximum payload mass fraction for the given variables.

52

3.5.4 Power.   For a thruster to operate it needs power. The jet power is the amount of power required for the the thruster to operate at a given thrust and specific impulse. The equation for the jet power is given by

$$P_J = \frac{m_{dot}v_e^2}{2} = \frac{Tv_e}{2} \ [15]$$ (3.19)

where

$$v_e = I_{sp}g_0$$ (3.20)

and m_{dot} is the mass flow in kg/s, T is thrust in newtons, and I_{sp} is the specific impulse.

However, since thrusters are not 100 percent efficient a thruster efficiency term, the same one from the payload mass fraction, is introduced into the jet power equation. This produces the equation for source power.

$$P_s = \frac{P_J}{\eta_t} = \frac{m_{dot}v_e^2}{2\eta_t} = \frac{Tv_e}{2\eta_t} \ [15]$$ (3.21)

As a result, higher thrust and higher specific impulses require correspondingly higher powers. Therefore, the thrust and specific impulse of the spacecraft will be limited by the power available.

3.5.5 Gravity Assist.   Looking back at Figure 3.9 and Figure 3.8, to reach 550 *AU* in approximately 90 years would require a ΔV in the 60-70 *km/sec* range. To decrease that requirement the spacecraft could make use of a gravity assist. Gravity assists utilize the one or multiple gravity wells to substantially increase the velocity of the spacecraft relative to the sun. Thus, the options available are: Jupiter Gravity Assist (JGA), Saturn Gravity Assist (SGA), Uranus Gravity Assist (UGA), Neptune Gravity Assist (NGA), Jupiter and Saturn Gravity Assist (JSGA), Jupiter and Uranus Gravity Assist (JUGA), Jupiter and Neptune Gravity Assist (JNGA),Solar Gravity Assist (SoGA), Venus and Solar Gravity Assist (VSGA). Furthermore, if the spacecraft is to be launched in a specific direction examining the sidereal period of the planets produces the time between launch opportunities, which is pictured in Figure 3.10.

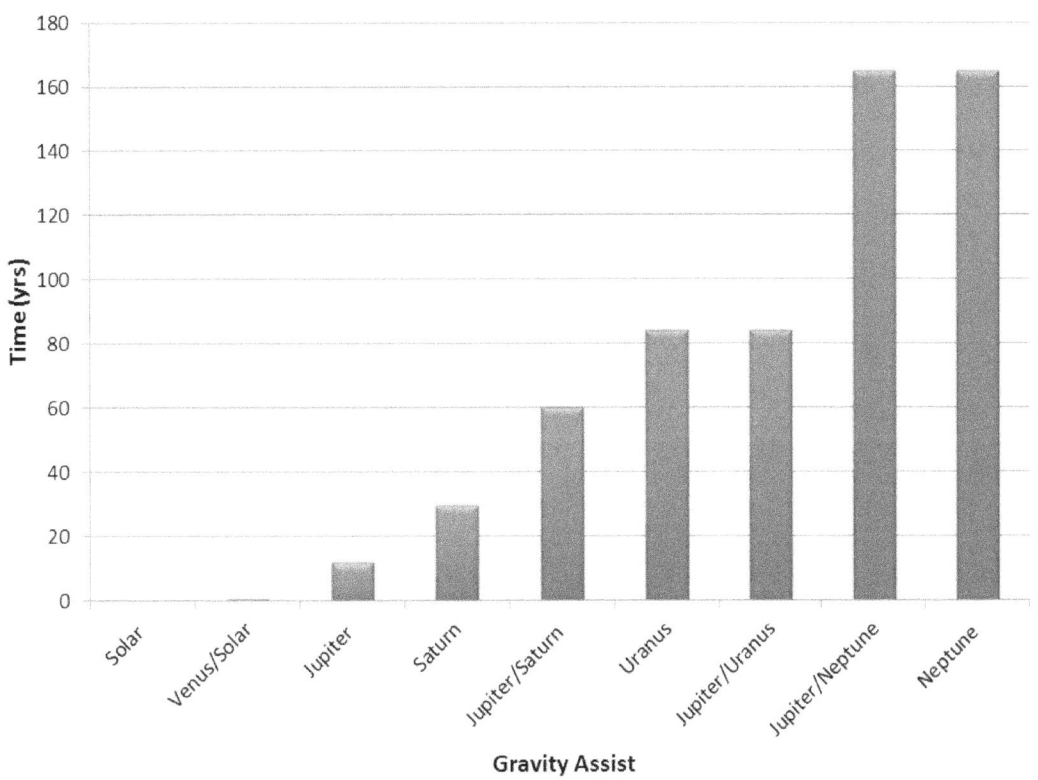

Figure 3.10: Time Between Launch Opportunities   [8] [9]

Figure 3.10 shows that the only plausible gravity assists are the SoGA, VSGA, JGA, and SGA. The others are unreasonable because the time between launch opportunities is too large, i.e. greater than 50 years. The SoGA and VSGA would require close approaches to the sun necessitating the use of large, heavy radiation shielding. Furthermore, the physics of gravity assist dictate that the more massive the planet the more velocity it provides; thus, since Saturn is much less massive than Jupiter out of the four remaining trajectories the most promising is the JGA.

The physics of gravity assists can be understood using analytic expressions. The following approach to gravity assists assumes an approximately coplanar flyby. Furthermore, the calculations for a gravity assist at Jupiter require input values for the radius of perigee, r_p; insertion velocity at "infinity", v_∞; mass of the planet, M_J; velocity of the planet, V_J; and the orientation angle, ϕ. The goal of the gravity assist calculation is to find how much ΔV the planet provides. The first required expression is that of the semi-major axis. The semi-major axis is defined as the largest radius of the orbit's conic section. This relation is stated as

$$a = -\frac{GM_J}{v_\infty^2} \quad [10] \tag{3.22}$$

where the only non-input variable is G the gravitational constant. Following the calculation of the semi-major axis is the eccentricity e, which measures the eccentricity of the gravity assist orbit.

$$e = 1 + \frac{r_p v_\infty^2}{GM_J} \quad [10] \tag{3.23}$$

In Equation 3.23, all of the variables have been previously defined. Next, the value of the parameter can be calculated with the eccentricity and the semi-major axis. The parameter is a property of conic sections dictated by the expression

$$p = a(1 - e^2) \quad [10] \tag{3.24}$$

After the parameter the next quantity of interest is the periapsis velocity v_p. The periapsis velocity is the velocity at the minimum orbital radius from Jupiter.

$$v_p = \sqrt{\frac{2GM_J}{r_p} + v_\infty^2} \quad [10] \tag{3.25}$$

This velocity can then be used to calculate the angular momentum h of the orbit.

$$h = r_p v_p \quad [10] \tag{3.26}$$

The values of Equations 3.22, 3.23, 3.24, 3.25, and 3.26 are all constants throughout the calculations. The first equation of note with a varying expression is the true anomaly.

The true anomaly measures the position of the spacecraft in the orbit relative to periapsis, it is negative prior to it and positive following it. Thus, periapsis has a true anomaly of zero. True anomaly is given by the expression

$$f = \cos^{-1}\left(\frac{\frac{p}{r} - 1}{e}\right) \; [10] \tag{3.27}$$

where r is the current radius of the gravity assist. Furthermore, to find the initial and final true anomaly examine Equation 3.27 and allow r to go to infinity, because the initial point and the final point of the gravity assist are an arbitrarily large distance away. As a result Equation 3.27 becomes

$$f_\infty = \cos^{-1}\left(-\frac{1}{e}\right) \; [10] \tag{3.28}$$

Following that, Equation 3.29 produces the current radius of the gravity assist at any point in the orbit.

$$r = \frac{p}{1 + e \cos f} \; [10] \tag{3.29}$$

Varying true anomaly from $-f_\infty$ to $+f_\infty$ creates an entire map of the radii from start to finish. These radii can then be inserted into Equation 3.30 to yield its corresponding velocity,

$$v = \sqrt{\frac{2GM_J}{r} + v_\infty^2} \; [10] \tag{3.30}$$

Next, the range angle β can be thought of as a measure of displacement from the initial true anomaly.

$$\beta = f_\infty + f \; [10] \tag{3.31}$$

The flight path angle γ is given by

$$\gamma = \mp \cos^{-1}\left(\frac{h}{rv}\right) \; [10] \tag{3.32}$$

where the radius and velocity of the orbit are constantly changing. The minus plus sign is present because while the true anomaly is negative the flight path angle is also negative,

56

and while the true anomaly is positive the flight path angle is positive. Then, combining the flight path angle and range angle into an equation produces the turn angle δ.

$$\delta = \beta - \gamma - 90° \quad [10] \tag{3.33}$$

The turn angle "is a measure of how much the spacecraft's velocity orientation has been rotated from its starting direction toward its final direction" [10]. Finally, Equation 3.34 builds on the knowledge provided by the previous equations to produce the spacecraft's velocity relative to the sun V.

$$V = \sqrt{v^2 + V_J^2 - 2vV_J \cos(\phi + \delta)} \quad [10] \tag{3.34}$$

Figure 3.11 demonstrates the effect of gravity assists very clearly and concisely. It shows the velocity profile of Voyager I's Jupiter gravity assist. Upon entering its Jupiter gravity assist maneuver Voyager I possessed a velocity of 12.62 km/sec relative to the Sun. At its departure from the maneuver Voyager I's velocity relative to the sun had increased by 10.77 km/sec to 23.39 km/sec. Nonetheless, Voyager I's velocity relative to Jupiter was the same on entry and departure.

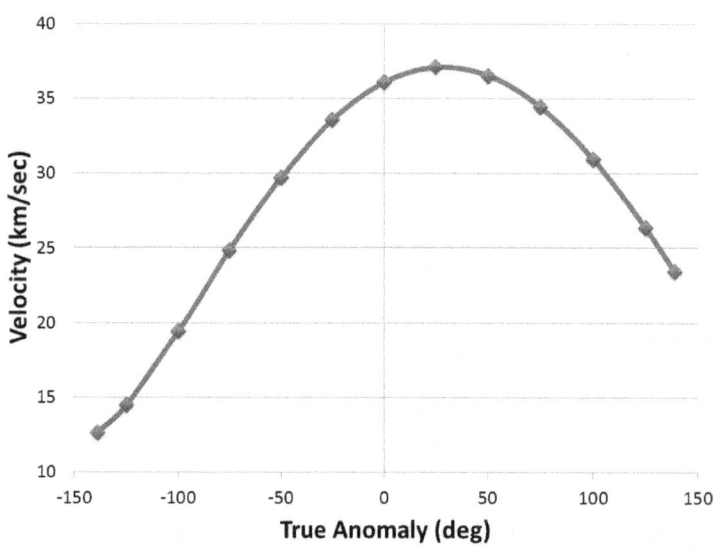

Figure 3.11: Voyager I Gravity Assist Velocity Relative to the Sun   [10]

3.6 Attitude Control

The attitude control subsystem is driven by the pointing requirements. For the FSM the pointing requirements are dictated not by the mission–pointing at the Sun–but by communications–pointing at the Earth. The basic pointing requirement equation can be expressed as

$$\psi = \frac{D}{h} \ [35] \tag{3.35}$$

where ψ is the pointing accuracy, D is the target diameter, and h is the distance from the spacecraft to the target. Interestingly, as the pointing requirement decreases it becomes harder to maintain and more accurate sensors are necessary. What is more, while h is explicitly stated D must be calculated. For the case of the FSM, the target diameter is the distance subtended by half the $HPBW$ at Earth, because with this tolerance the Earth will remain within the $HPBW$.

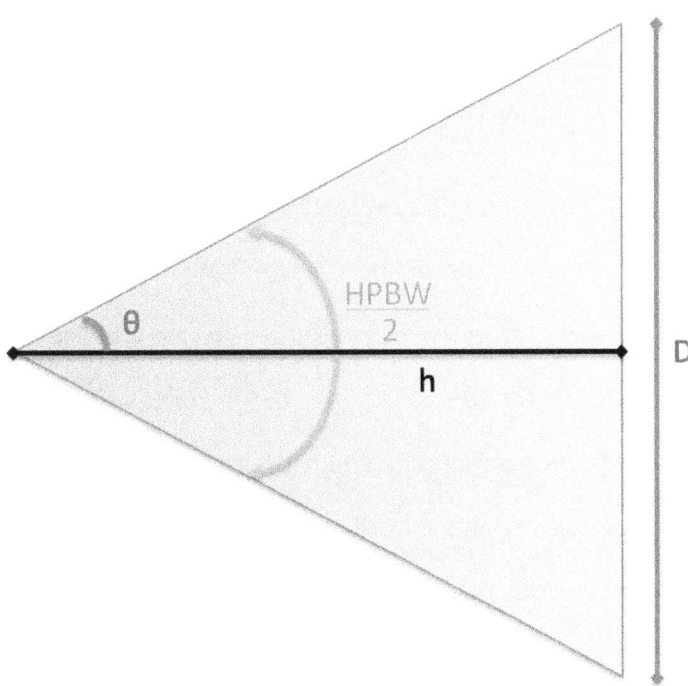

Figure 3.12: Pointing Accuracy Geometry

The geometry of the pointing requirement is simple. At the apex of Figure 3.12 is the spacecraft transmitter. The straight line distance between the current location of the spacecraft and Earth is represented by h. The half of the HPBW, which was mentioned earlier in the Communications and Data Handling subsection on page 45, subtends the target diameter at Earth. Finally, θ is a quarter of the HPBW and was introduced to simplify solving for the target diameter.

Looking at Figure 3.12 the relationship between θ and D can be stated as

$$D = 2h \tan(\theta) \tag{3.36}$$

Thus the target diameter increases with the $HPBW$. By inserting Equation 3.11 in for θ Equation 3.36 becomes

$$D = 2h \tan\left(\frac{HPBW}{4}\right) = 2h \tan\left(\frac{21}{4 f_{GHz} D_{ant}}\right) \tag{3.37}$$

Equation 3.37 illustrates the inverse relationship between the target diameter and the antenna diameter. Furthermore, plugging Equation 3.37 back into Equation 3.35 produces

$$\psi = 2 \tan\left(\frac{21}{4 f_{GHz} D_{ant}}\right) \tag{3.38}$$

which is the pointing requirement equation in its simplest form. Solving this equation using the specifics for the FSM produces a pointing accuracy of approximately 20 degrees; however, by limiting the pointing accuracy to five degrees the data rate will see a requisite increase due to smaller pointing losses. Finally, to achieve the *specified pointing accuracy of five degrees* the spacecraft needs a robust attitude control sensor and actuator suite.

Table 2.2 displays the attitude sensors and actuators used on NASA's deep space missions. The sensors include star trackers, sun sensors, and inertial reference units. Sun sensors track the spacecraft's attitude by knowing its own position on the spacecraft reference frame and comparing that with the sun vector it measures. Unfortunately, sun senors can only track two attitude dimensions whether that be pitch and yaw, pitch and roll, or yaw and roll [35]. Star trackers, on the other hand, can target multiple stars and produce information on all three attitude dimensions. Finally, inertial reference units do not require an external reference. They are capable of measuring torques on the spacecraft using well-calibrated gyroscopes and accelerometers. The measured outputs from the gyroscopes and accelerometers are then translated into information about the spacecraft's attitude. The FSM will use only two of the sensors to ensure that it can achieve the pointing accuracy required to complete the mission. A sun sensor will not be used because operating at the mission distance the sun will appear to be just another star. Knowing the spacecraft attitude is not enough. The FSM must also be able to control its attitude.

The primary active deep space attitude actuators are thrusters. Thrusters are a reliable solution for long term attitude actuation. They were chosen over the inclusion of CMGs and reaction wheels because while CMGs and reaction wheels would decrease the need for propellant they have a much higher propensity to fail on a long duration mission according to Larson and Wertz's text [15]. The principal thruster propellant is MMH/N_2O_4, because of its desirable specific impulse and ability to be stored for long durations. Thus thrusters will be used as the active actuator on the FSM. Of course, thrusters require propellant which, in turn, require propellant tanks.

The amount of propellant required is determined from the Ideal Rocket Equation, Equation 3.3. In this case, the known values are the initial mass, specific impulse, and ΔV since the amount of ΔV provided by the thruster is predetermined based on an expectation of disturbance torques. And the desired value is the final mass. Once the final mass is found the propellant mass is the difference between the initial and final mass.

The propellant tank size is dependant upon the tank material and volume of the propellant. For the FSM the tank material will be Aluminum-2219, because of its ultimate strength and density. Aluminum-2219 has an ultimate strength of .413 GPa and density of

$2800\ kg/m^3$ [38]. Moreover, it is compatible with both MMH and N_2O_4 [38]. Nitrogen Tetroxide and MMH have densities of $1440\ kg/m^3$ and $878\ kg/m^3$ respectively [38].

After using Equation 3.3 to calculate the amount of total propellant required the total propellant mass is partitioned based on the desired fuel to oxidizer mass ratio. Then, the following equation is used to calculate the radius of the spherical tank.

$$r_s = \sqrt[3]{\frac{3(1.1)V_p}{4\pi}}\ \text{[38]} \tag{3.39}$$

where V_p is the volume of the propellant. The constant value of 1.1 takes care of the ullage and trapped volumes. The ullage and trapped volumes account for the fact that the tanks will not be completely full with propellant but will leave space so that there will be enough pressure to push the final propellant through the pipes. In this case a conservative ten percent is left empty in the tank. Next, Equation 3.40 calculates the area of the tank using the radius.

$$A_s = 4\pi r_s^2\ \text{[38]} \tag{3.40}$$

The tank wall thickness can be calculated using the radius along with the burst pressure and ultimate strength.

$$t_s = \frac{p_b r_s}{2F_u}\ \text{[38]} \tag{3.41}$$

where

$$p_b = 2P_{ME} \tag{3.42}$$

In Equation 3.42, P_{ME} is the maximum expected operating pressure. For both the fuel and the oxidizer the maximum operating pressure will be around $1.7\ MPa$ so that the pressure in the tank is sufficient to meet the demands of a bipropellant thruster. The constant 2 in the equation is the factor of safety. Finally, the mass of the tank can be calculated through

$$m_s = A_s t_s \rho_{mat}\ \text{[38]} \tag{3.43}$$

where ρ_{mat} is the density of the tank material. Outside of the active actuation of thrusters there are also passive attitude actuators.

For deep space missions there are only two viable passive attitude actuators–nutation dampers and spin stabilization. Nutation dampers absorb torques on a given axis and convert them into heat. The second form of passive actuation, spin stabilization, uses "the conservation of angular momentum to maintain a constant inertial orientation of one of its axes" [35]. Thus the spacecraft can remain pointed toward Earth or the Sun and be slewed by the thrusters when necessary. Spin stabilization cannot be used on the FSM during the boost because of the need to keep the solar panels pointed at the sun.

3.7 Environmental Control

The environment of the FSM will not be decidedly different then that of previous deep space missions. There are two principal environmental concerns. They are thermal balance and irradiation.

Maintaining thermal balance means having the the amount of heat coming into the spacecraft equal the amount of heat being radiated away from the spacecraft, which is mathematically illustrated by the following equation

$$Q_{in} = Q_{out} \tag{3.44}$$

where heat is designated by the quantity Q. This is the steady state thermal balance equation. The quantity on the left side of the expression includes heat generated by the spacecraft and imparted on the spacecraft by planets and the sun. The right side accounts for heat radiated away by passive or active means. Active means includes radiators, louvers, heaters, and heat pipes. Passive means include blankets and reflective materials. The goal of thermal balance is to keep the spacecraft temperature in a desirable range. Table 3.2 provides typical thermal ranges for spacecraft components. The outer

spacecraft temperatures can range between ±200°C [15]. Accordingly, a significant amount of thermal protection is needed.

Table 3.2: Typical Thermal Ranges for Spacecraft Components   [15]

Component	Typical Temperature Range (°C)	
	Operational	Survival
Inertial Reference Unit	0 – 40	-10 – 50
Star Trackers	0 – 30	-10 – 40
C & DH	-20 – 60	-40 – 75
Hydrazine Tanks	15 – 40	5 – 50
Antennas	-100 – 100	-120 – 120
Solar Panels	-150 – 110	-200 – 130
Total Range	15 – 30	5 – 40

Radioisotope heaters units (*RHU*) will be used as an active means to maintain the specific temperature range of each component. These devices, discussed in the Background and Theory, radiate heat from the decay of radioisotopes. Furthermore, the FSM will make use of low-emittance aluminized mylar multi-layered insulation blankets to prevent excessive heating/heat loss.

Not only must thermal balance be maintained but heat must also distributed around the spacecraft. A high concentration of heat in one area or the lack heat in another would be detrimental to the spacecraft. However, heat pipes will not be used to distribute heat throughout the spacecraft. Heat will be distributed by thermally tying the spacecraft together.

Like the thermal environment the radiation environment will closely match that of previous space missions therefore previous radiation hardening schemes can be used. The novel threat the spacecraft will face is prolonged exposure to the same rate of irradiation. This will increase the likelihood of a malfunction. Nonetheless providing protection against irradiation is a simple matter. Either dense materials must be introduced to stop the high energy particles from reaching the electronics or high reflectance materials to reflect particles away from the spacecraft. There are also internal radiation threats to the spacecraft due to the use of RTGs. As a result, the RTGs must be placed to minimize their effect on the primary payload.

3.8 Conclusion

In conclusion this chapter analyzed and discussed how to optimize the gravitational lens mission. The following chapter will utilize the findings of this chapter to design the FSM.

4 Design

4.1 Design Flow

The spacecraft requirements have been presented; thus, the purpose of this chapter is to synthesize the outcomes of the previous chapters and the spacecraft requirements into one coherent spacecraft design through the iterative process shown in the Design Flow Diagram, Figure 4.1.

At the top of the diagram is the propulsion system. As stated in Section 3.2, the type of propulsion system, high specific impulse or low specific impulse, decided which launch vehicle is utilized. What is more, the propulsion system also has a direct effect on the ΔV. It is for these two reasons that the propulsion system was chosen as the starting point of the iterative process.

Under the propulsion system is the launch vehicle. For this analysis, the launch vehicle was directly dependent on only the propulsion subsytem. The launch vehicle then puts limits on the mass and volume of the entire spacecraft. Progressing to the next level, every spacecraft subsystem was interdependent.

In conclusion, there were several iterations through the design flow due to the hierarchy and coupling of dependencies. The following sections present the final result of these iterations.

Figure 4.1: Design Flow Diagram

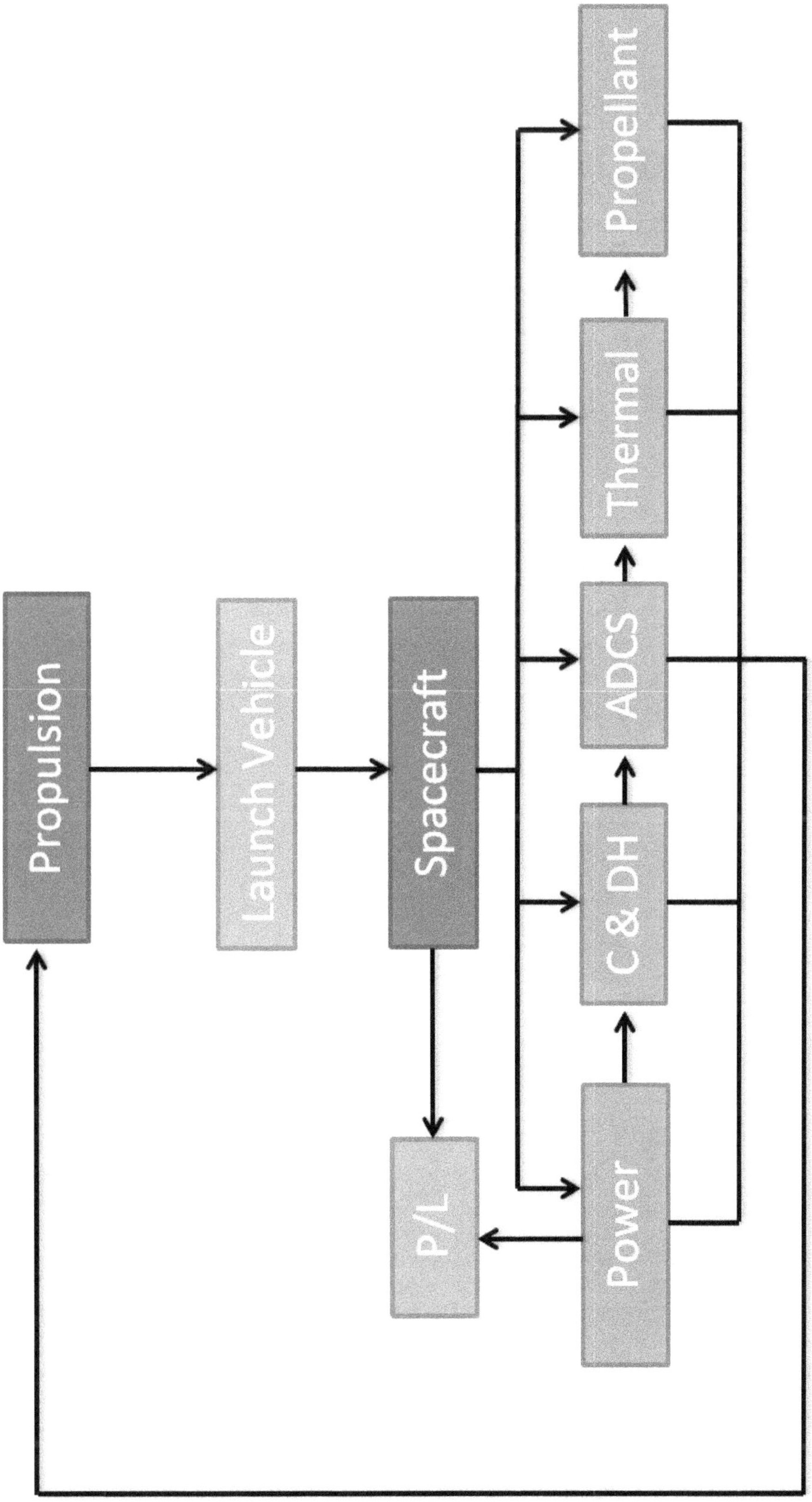

4.2 Launch Vehicle

The analysis in Chapter III stipulated that when a low thrust propulsion system is present the optimal launch vehicle choice is the one that provides the largest delivered mass to the parabolic excess energy. Therefore, from the chart of options on pg. 35, the obvious choice is the Delta IV-H/Star 48/Star 37 configuration. This provides a delivered mass of 8608 kg; thus, limiting the total spacecraft mass to 8608 kg.

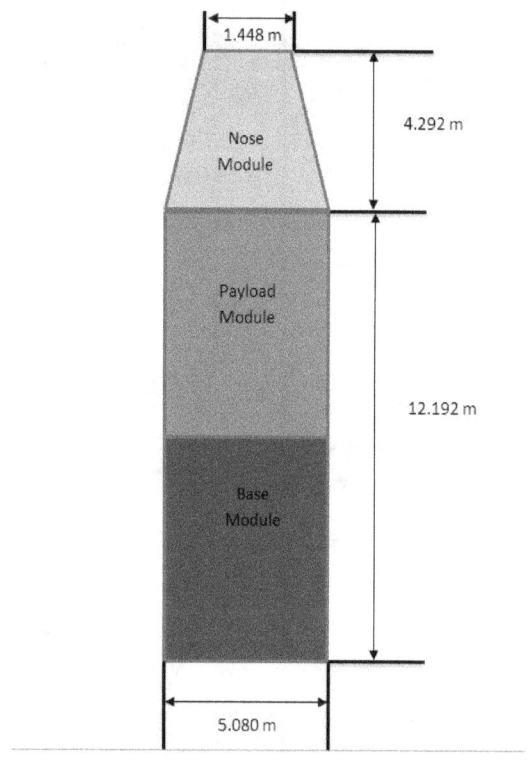

Figure 4.2: Payload Fairing [11]

With that, the launch vehicle's payload fairing constrains the dimensions of the spacecraft. Figure 4.2 displays the standard Delta IV metallic payload fairing used by the United States government. Clearly from this illustration the majority of spacecraft must be less than 5.08 meters in diameter and 12.19 meters in height in its stowed configuration. Or if it is rectangular it must have sides that measure less than 3.59 meters so that it will fit inside the circular envelope. Finally, The spacecraft may have appendages that extend up to 4.29 meters from the apex of the main spacecraft.

4.3 Spacecraft

4.3.1 Payload.   The primary payload is a 12-meter diameter parabolic center-feed antenna/radio telescope. It is this size so that it can accurately measure variations in the frequency magnification, detect possible communications sent through the gravitational lens, and take images images using radio telescopy.

The dual purpose primary payload will be a deployable metallic mesh similar to Galileo and the Tracking and Data Relay Satellites [42]. The metallic mesh has an approximate specific mass of .12 kg/m^2. With an area of 113.10 m^2 the antenna mass will be 13.5 kg. The antenna mass must also account for the mass of the feed array structure, which is approximately 10 kg [15]. Therefore, the total mass of the primary payload is 23.5 kg. The power seen by the primary payload will vary with the degradation of the power source, but the goal is to maintain 15 W. The secondary payloads were outlined in the Background and Theory. Figure 4.1 consolidates the mass and power requirements of the payloads.

4.3.2 Communications and Data Handling.   **The primary payload will also serve as the high gain antenna.** Accompanying it will be a medium gain parabolic antenna whose mission it will be to acquire Earth and to maintain beacon monitoring between the spacecraft and the DSN, located in California, USA; Madrid, Spain; and Canberra, Australia.

Using Equation 3.6 the gain of the primary HGA is 46.59 dB. Since the goal of the MGA is to acquire Earth it is desirable for it to have a high *HPBW*. Furthermore, according to Equation 3.11 to do that the diameter of the antenna needs to be minimized. In contrast, the gain equation suggests a large diameter. As a result, the chosen MGA

Table 4.1: Payload Performance   [15]

Payload	Mission	Mass (kg)	Data Rate (Mbps)	Power (W)	Attitude Requirements	Section
VBSDC	Dust Detection	1.6	0.02	5.1	-	2.3
FGM	Magnetic Field Measurements	15.25	0.03	11.3	-	2.3
SHM	Magnetic Field Measurements	9.08	0.01	6.5	-	2.3
PEPSSI	Detect Neutral and Ionized Particles	1.5	0.01	2.5	-	2.3
Radio Telescope	Characterize Gravitational Lens	23.5	2	15	5°	4.3.1
TOTAL		50.93	2.07	40.4		

diameter is 2-meters in diameter. This choice balances the need for a respectable gain in deep space and the desire for a higher *HPBW*. The MGA, according to its area and the specific mass of metallic mesh, weighs .38 kg. Concurrently, the array structure for the MGA weighs 5 kg. Its area also corresponds to a *HPBW* of 4.58°, which does not sound large but, in fact, it is six times larger than the HGA *HPBW* of 0.7636°. The gain of the MGA is 31.027 dB. The corresponding data rates will be addressed in Section 4.5.

Finally, in regard to data handling, the processor will be the RAD750 and the FSM will use a 64 Gbit solid state recorder as stated in Chapter III. The data handling and processing functions of the spacecraft are expected to draw approximately 15 W [15]. Figure 4.3 shows the payload/communications architecture for the FSM.

4.3.3 Power.   The ASRG has only attained TRL 6 thus it cannot be considered COTS. As a result, the flight proven GPHS was chosen based on its superior power density compared to the other options in Table 2.4. The GPHS is pictured in Figure 4.4.

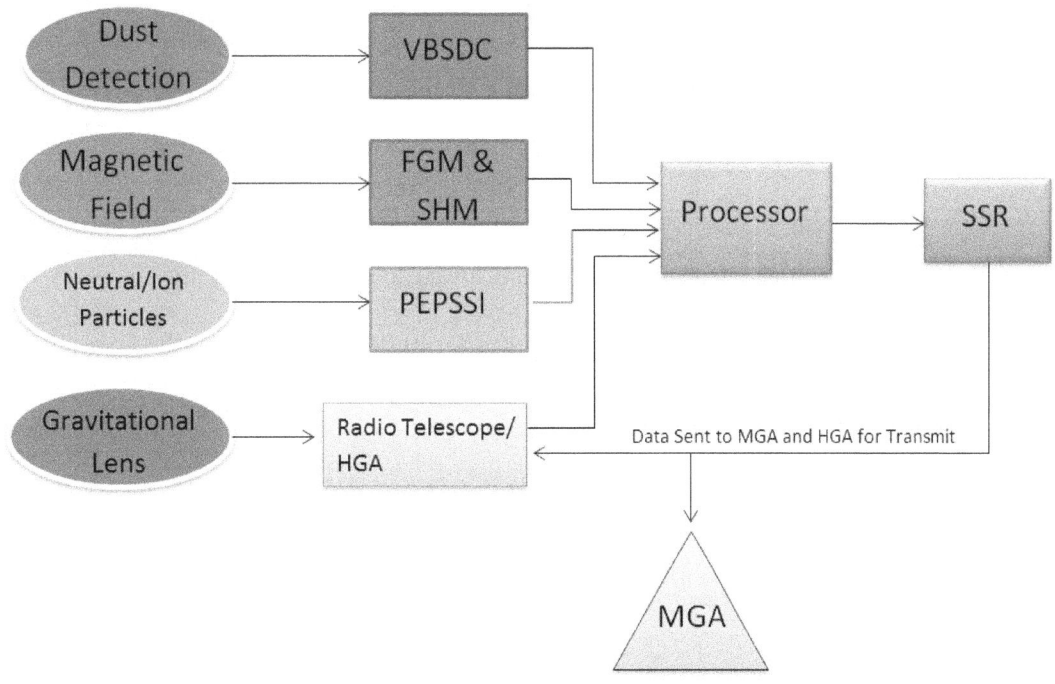

Figure 4.3: Payload and Communications Architecture. Note: The primary payload serves as a radio telescope and HGA.

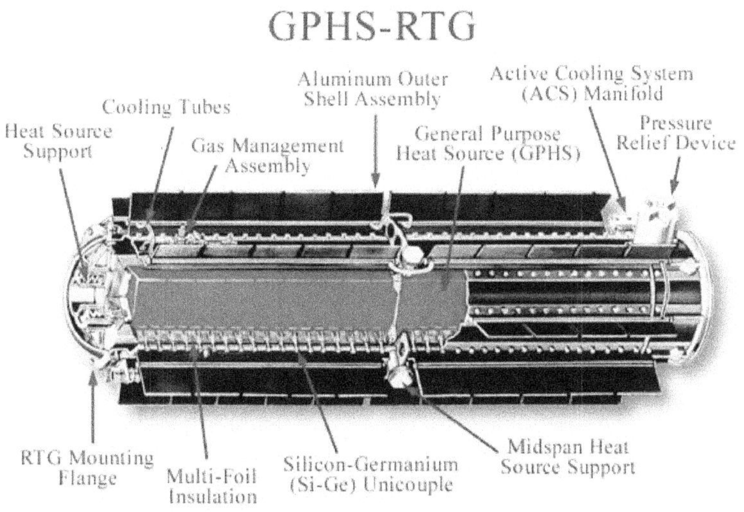

GPHS-RTG

Figure 4.4: General Purpose Heat Source RTG   [12]

The number of generators required is usually determined by the EOL requirements, which were taken into account, but for the FSM the power requirement dictated by the thruster superseded that of the EOL. Thus the beginning of life generator power level is 4914 W so that thrust can be maintained after jettisoning the boost phase at Jupiter. This number was found by trading the mass of the power system, propellant mass, trip time, the power provided by the power system, and the propulsion levels. Figure 4.5 displays the result. Therefore, the optimum number of GPHS RTGs for the FSM is 20, because it corresponds to the minimum trip time.

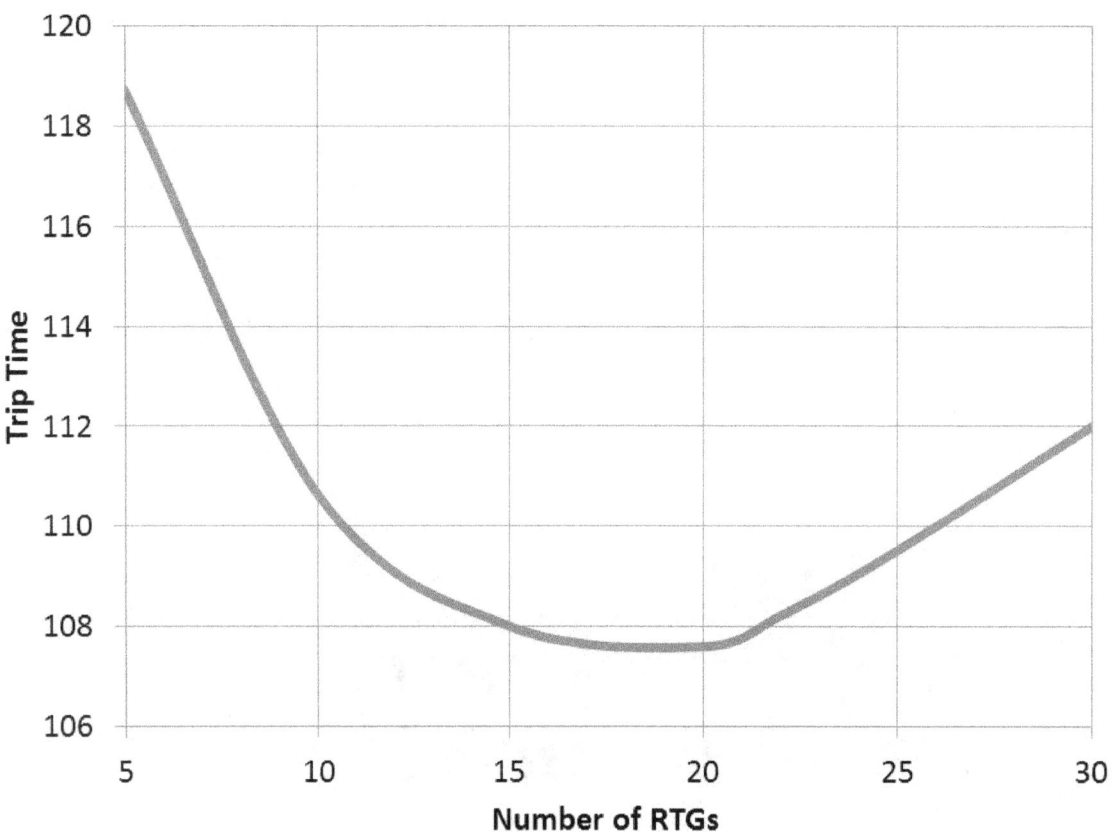

Figure 4.5: Power Source Trade

This corresponds to an EOL power of approximately 1184 W at 1000 AU, which is more than sufficient for mission operations based on the total power draw of the spacecraft presented in Table 4.11.

At 57.8 kg per GPHS the total weight of the GPHS power system is 1156 kg, which requires 156 kg of plutonium-238 . This presents an intractable problem that will be discussed in Chapter V. The stellar performance demonstrated by its forbearers on Voyager and Pioneer imply that these generators can operate for extremely long durations. Finally, Figure 4.14 displays the expected degradation of the power source as well as the power draw.

The second challenge for the power system of the FSM is to provide sufficient power for the boost phase prior to the JGA. Analysis of this problem showed that the optimal power strategy would be to utilize a solar electric phase. Furthermore, the solar array should be able to provide sufficient power for the thruster throttling profile. A good candidate is the Spectrolab Ultra Triple Junction $GaInP_2/GaAs/Ge$ solar cells. A 60-m^2, 900 kg optimized array of these cells would theoretically produce approximately 22 kW of power at air mass zero (AM0) [38] [31]. That may seem heavy but it would take more than 5000 kg of this mission's RTGs to produce a similar power.

The cells have a 28% BOL efficiency, and a 24.3% EOL efficiency, where EOL is defined as after 1 MeV electron fluence of $1 \times 10^{15} \frac{e}{cm^2}$ [31]. Also, since solar arrays are sensitive to the sun angle the attitude control system will keep the FSM's solar array pointed toward the sun. Finally, as the efficiency degrades and the solar panels travel away from their optimal range the power will drop off steeply; hence, when the spacecraft reaches Jupiter the solar stage will be jettisoned. Table 4.2 consolidates the information presented in this section.

Table 4.2: Power Performance

	BOL Power (kW)	Total Dimensions (m)	Total Mass (kg)
ASRG	4.91	.422 x 1.14	1156
Solar Panels	22	3 x 20	900
Total	26.91	-	2056

4.3.4 Propulsion. The NASA Evolutionary Xenon Thruster NEXT is by far the best thruster choice. It has records for the most hours of operation, highest propellant throughput, greatest total impulse demonstrated, and longest hollow cathode operation [43]. More specifically, it has demonstrated the ability to process greater 632 kg of Xenon over 4.2 years.

The NEXT thruster can be throttled through its forty throttle levels from 6.9 kW down to .5 kW [43]. This corresponds to a specific impulse range from approximately 4300 sec to 1480 sec and thrust range of 25.4 mN to 234.6 mN [43][44]. Table 4.3 shows the sizing and power requirements of the thruster while Figure 4.6 displays the entire thruster system.

Table 4.3: NEXT Sizing and Power [13]

	Mass (kg)	Dimensions (cm)	Power (W)
Thruster	12.7	Diameter – 55 Length - 44	6900 – 500
Power Processing Unit	33.9	42 x 53 x 14	N/A
Propellant Management System	5	*High Pressure Assembly* 33x15x6.4 *Low Pressure Assembly* 38x30x6.4	*High Pressure Assembly* 4.3 *Low Pressure Assembly* 15.9
Gimbal	6	61x72	N/A
Total	57.6		

Thus far NEXT has processed 632 kg of Xenon in the course of the long duration testing. Yet it has demonstrated only slightly decreasing efficiency. Whereas NSTAR,

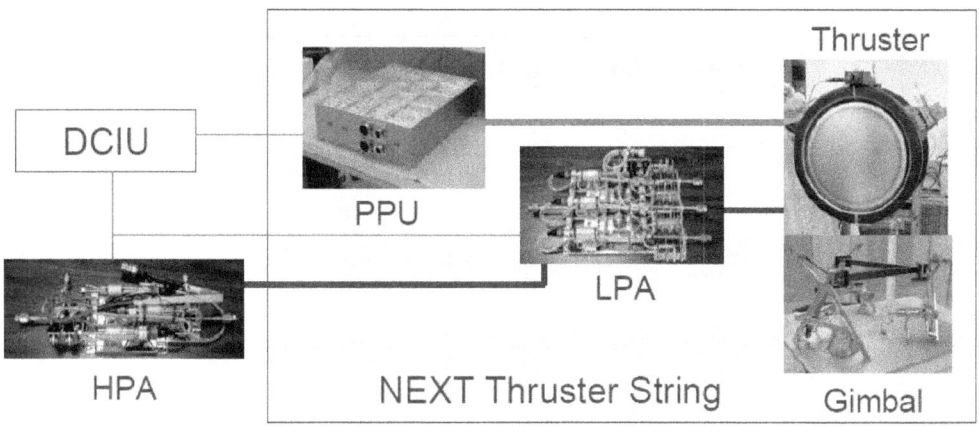

Figure 4.6: NEXT Thruster System [13]

NEXT's predecessor, lost 9% efficiency over the same period NEXT has lost only 2.5% engine efficiency [43]. Furthermore, NEXT is forecasted to be capable of processing 800 kg at the maximum power level, failing primarily due to accelerator grid erosion, which is the result of charge-exchange erosion [43]. Charge-exchange erosion is based on propellant throughput and input powers meaning the thruster is life limited by the amount of propellant processed.

The analysis which was used to chose the number of power sources also determined the propellant load of 4885 kg of xenon. This is what remains of the delivered mass after the spacecraft dry mass–2514 kg–and the boost phase dry mass–1200 kg–are taken into account. Then, the required number of thrusters was derived from the propellant mass by taking the total propellant mass and dividing it by 640 kg, which is the predicted maximum throughput minus twenty percent for margin. The resulting number of thrusters is 7.63 thrusters. Hence, a conservative eight thrusters are required to process that amount of propellant. Meaning that each thruster operating one at a time will process, on average, 610 kg of Xenon. Eight thrusters require the components listed in Table 4.4. Finally, Equation 4.1 displays the mass of the xenon tank as a function of the mass of xenon propellant.

$$m_{Xe\ tank} = 52 + 0.075m_{prop} + .154m_{prop}^{\frac{2}{3}} \quad [38] \tag{4.1}$$

The arrangement of the spacecraft and the reasons for it will be discussed in more detail in Section 4.5.3 and Section 4.4, but at this moment it is important to note that the spacecraft will be split into two segments. The first segment will be a solar electric boost stage and the second segment will be the main spacecraft. It is important to note at this point because the xenon will be split into two tanks–one in the boost phase and one in the solar electric phase–to produce a higher total ΔV. Therefore, using Equation 4.1, the mass of the first tank is 308.7 kg and the second is 217.13 kg, which correspond to propellant throughputs of 2996 kg and 1888 kg respectively.

Table 4.4: NEXT Component List [13]

	Number of Components	Mass	Power (W)
HPA	1	1.9	1.9
LPA	4	12.4	3.1
PPU	4	135.6	610 - 7220
Gimbal	4	24	-
Thruster	4	50.8	500 – 6900
Total	17	224.7	615-7225

4.3.5 Attitude.

4.3.5.1 Attitude Determination. Attitude determination will be managed by star trackers and an inertial reference unit. This section will present a COTS component that meets the mission requirement.

The Ball Aerospace CT-633 will serve as the COTS star tracker. It is capable of tracking up to five stars and providing an angular accuracy of 25 arc seconds, temporal accuracy of 6.5 arc seconds, and attitude accuracy of 15 arc seconds at the end of life [45].

75

The CT-633 has a power draw of 9 W, mass of 2.49 kg and dimensions measuring .135 m in diameter by .142 m in length [45].

The inertial reference unit chosen for the FSM is the Honeywell Miniature Inertial Measurement Unit ($MIMU$). It can sense attitude changes in all three axes in the range of ±375 deg/sec [46]. It has a mass of 4.44 kg, power consumption of 22 W, and measures .233 m in diameter by .169 m in height [46].

Table 4.5: Attitude Determination Mass and Power

	Mass (kg)	Power (W)
Star Tracker	2.49	9
MIMU	4.4	22
Total	6.89	31

4.3.5.2 Attitude Control.   The active attitude actuation of the spacecraft will be handled by twelve .8 N N_2O_4/MMH thrusters. At a 1-to-1 mixture ratio the specific impulse of the thruster is 318 sec [38]. Furthermore, similar thrusters were used on both of the Voyager missions and New Horizons. The New Horizons qualification testing found that thrusters are reliable up to at least 400,000 cycles [5]. On the Voyager mission these thrusters lasted for more than 500,000 cycles [5]. Moreover, the FSM attitude control profile requires 5 corrections a month. Over two hundred years that corresponds to 240,000 cycles.

Similar thrusters weigh in a range of .2 kg to .4 kg each [47]. Thus, for the FSM each thruster will be assumed to have a mass of .4 kg and dimensions of .4 m in diameter and .15 m in height. Also, the catalyst bed for each thruster will require 2.2 W of power and the entire thruster system can draw up to 50 W [15].

Thruster performance was examined by looking at the slew rate of the spacecraft in two different configurations. The first configuration was with the boost phase attached and the second configuration was with the boost phase removed. The first step in this analysis was to calculate the mass moments of inertia for the spacecraft. This was done by using the principles of superposition and the parallel axis theorem. Some basic assumptions were necessary.

Two assumptions were made for the mass moment of inertia analysis. The first assumption of the analysis was that the spacecraft and its appendages could accurately be modeled as simple three-dimensional shapes, e.g. rectangular prisms and cylinders. The second assumption of the analysis was that the mass was distributed evenly throughout these sections so that the center of mass of each of these shapes was also the centroid.

Table shows the calculated moments of inertia for the spacecraft. The axes of the spacecraft are arranged such that the x-axis is parallel to the face of the solar panels, the z-axis goes traverses the spacecraft from bottom to top and the y-axis completes the right-hand rule. The moments of inertia for the spacecraft with the boost phase are much larger than those for the spacecraft without the boost phase, which is to be expected due to the size of the solar panels, the body of the boost phase, and amount of propellant used between the two analysis points.

Table 4.6: FSM Mass Moments of Inertia

Axis	With Boost Phase (kg-m²)	Without Boost Phase (kg-m²)
X-axis	46190	1782
Z-axis	14011	5555
Y-axis	16174	3414

Then the slewing rates, Equation 4.2, were calculated using the moments of inertia I, the force of the thrusters F, the thruster burn time t, and the thruster's distance from the center of mass L.

$$\dot{\Theta} = \frac{Ft}{IL} \quad [15] \qquad\qquad (4.2)$$

The results are shown in Table 4.7. For the spacecraft with its boost phase the thrust time was assumed to be 5 seconds and for the spacecraft sans the boost phase the thrust time was assumed to be 1 second. This difference is to aid the thrusters with the large variations in the size of the mass moments of inertia. The distance from the thruster to the center of mass for both cases is approximately 1.5 meters since the diameter and height of the main spacecraft body is 3 meters.

Table 4.7: FSM Slew Rate

Axis	With Boost Phase (deg/sec)	Without Boost Phase (deg/sec)
X-axis	.003	.017
Y-axis	.011	.0055
Z-axis	.009	.009

Based on the number of thruster cycles the FSM disturbance environment is comparable to previous missions, e.g. New Horizons, so the thrusters will provide a similar 130 m/s of ΔV for margin (unexpected and disturbance) torques and 200 m/s of ΔV for alignment and slewing to meet the 5 degree pointing requirement of the communications system. This corresponds to a total propellant mass of 150.72 kg with 75.36 kg of N_2O_4 and 75.36 kg of MMH. Table 4.8 was created using the propellant masses and equations from Section 3.6.

Attitude thrusters also require a propellant pressure system to pump the propellant out of the chamber and into the thrusters. The propellant pressure system on the FSM will have a mass of 10 kg. In conclusion, Figure 4.7 diagrams the attitude control subsystem on the FSM.

Table 4.8: Propellant Tank Design

	N$_2$O$_4$	MMH
Propellant Mass (kg)	75.36	75.36
Propellant Volume (m³)	.0523	.0858
Burst Pressure (Pa)	1.72 x 10^6	1.72 x 10^6
Tank Radius (m)	.2395	.2824
Tank Area (m²)	.7208	1.003
Tank Thickness (m)	4.99 x 10^{-4}	5.88 x 10^{-4}
Tank Mass (kg)	1.01	1.65

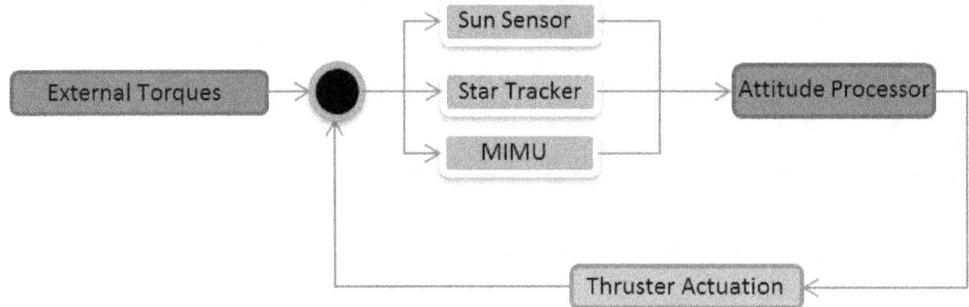

Figure 4.7: Attitude Control Loop

4.3.6 *Environmental Control.* Environmental control is one of the simplest subsystems. Furthermore, since the environment of the FSM is not decidedly different than that of previous space missions the majority of this subsystem will be carried over from those aforementioned missions. Nonetheless, the environmental control subsystem still serves as a multifaceted defense against the space environment .

Its first line of defense is on the exterior of the spacecraft in the form of MLI blankets. The number of layers of insulation was determined to be forty. This number was derived from the steady state heat transfer analysis based on the following equation which was itself derived from the thermal balance equation,

$$\epsilon = \frac{Q_{in}}{\sigma A(T_{s/c}^4 - T_\infty^4)} \text{ [15]} \qquad (4.3)$$

In Equation 4.3 , ϵ is the emissivity of the spacecraft, Q_{in} is the internally generated heat plus the solar flux, σ is the Stefan-Boltzmann constant, A is the exterior area of the spacecraft, $T_{s/c}$ is the temperature of the spacecraft, and T_∞ is the temperature of the surroundings. Furthermore, the spacecraft temperature was assumed to be 295 K since that is the mean acceptable operating temperature for electronics as defined by Table 3.2 and the temperature of the surroundings was assumed to be 2.7 K.

The analysis examined four different points in the spacecraft's trajectory: spacecraft at Earth with thruster on, spacecraft at Jupiter with thrusters off, spacecraft post-Jupiter with thrusters on, and spacecraft after thruster burnout. The internal heat generation was varied based on these four points show in Figure 4.9 . The spacecraft generates the least heat in the fourth case when the spacecraft thrusters have finished the thrust profile and the solar flux has declined dramatically. As a result, the emissivity required to maintain thermal balance came out to be the lowest at this point. More specifically, the emissivity required was .0020. That correlates to approximately forty layers of MLI [15]. At that thickness the multi-layered insulation blankets to be used on this mission have a specific mass of .73 kg/m^2 [15]. Accordingly, the total mass of the MLI blankets is 41.61 kg since the spacecraft main body is 57 m^2.

Since the number of layers of MLI is defined based on the coldest case radiators are needed to raise the emissivity when the spacecraft heat generation is higher. When the spacecraft is close to the Sun and firing thrusters it will be the warmest, and at that temperature the emissivity required to maintain thermal balance is .26. This is within the capability of COTS radiators. Thus the flight qualified radiator louver assembly will have a maximum emittance of .88 to bring the total emittance of the spacecraft up to .26 from

Table 4.9: Spacecraft Heat Balance

Condition	Q_{in} (kW)	Emissivity
Thrusting at Earth	6.5	.2655
Jupiter w/ Engine Off	.839	.0343
Post Jupiter Thrusting	1.928	.0787
Post Burnout	.051	.0020

the radiator off/louver closed emittance of .20. The radiator louver assembly has dimensions of 45.7 cm by 58.2 cm, an approximate specific mass of 4.5 kg/m^2 , and will draw up to 20 W[15].

While MLI blankets and radiator louver assemblies will maintain the temperature in the main spacecraft the FSM will use seven of the RHUs described in the Background & Theory to maintain the thermal balance of external sensors.

The final and most important line of defense against the space environment is radiation hardening. As a result, the spacecraft bus will be encapsulated in a tantalum radiation vault. Tantalum has a density of 16.69 $\frac{g}{cm^3}$, which means it is almost fifty percent more dense than lead. This will keep the radiation level the bus sees under the 100 krad mission requirement. The vault will be 0.35 cm in thickness and have dimensions of .75 × .75 × .75 meters. Thus, the vault will be 189 kg.

4.4 Model

This section displays models of the spacecraft design.

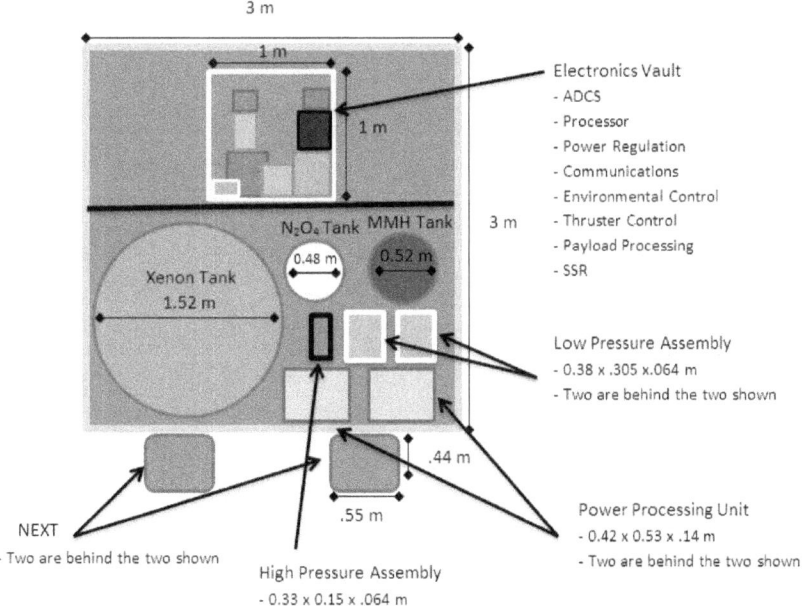

Figure 4.8: Internal Configuration

- Helium pressurant present in Xenon, N_2O_4 and MMH tanks

- All items fit within the 3 x 3 meter spacecraft body

- Width was necessitated by propellant tanks

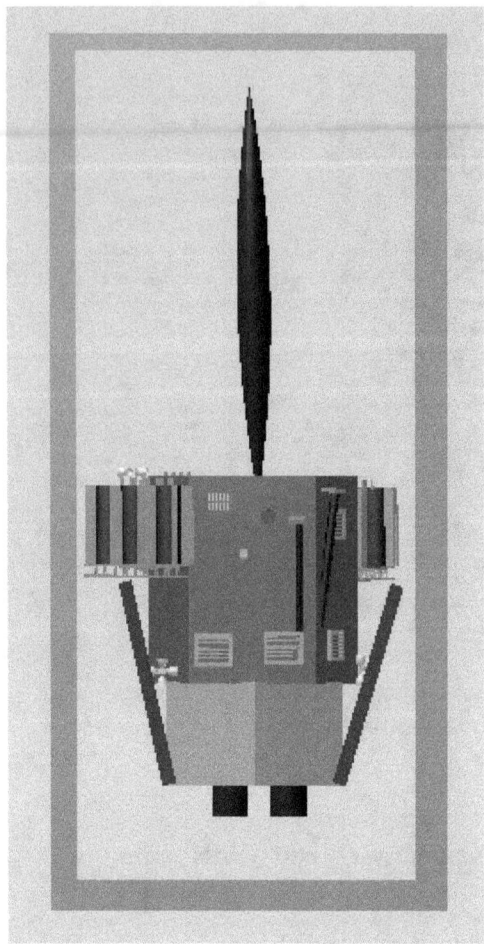

Figure 4.9: Stowed Configuration

- Shows the stowed configuration

- Fits within the payload and base module of the payload fairing

    Payload fairing shown in pink

    10.56 meters tall

    4.76 meters at widest

- HGA/Radio Antenna wrapped around antenna stem

- Solar panel folded in thirds and retracted

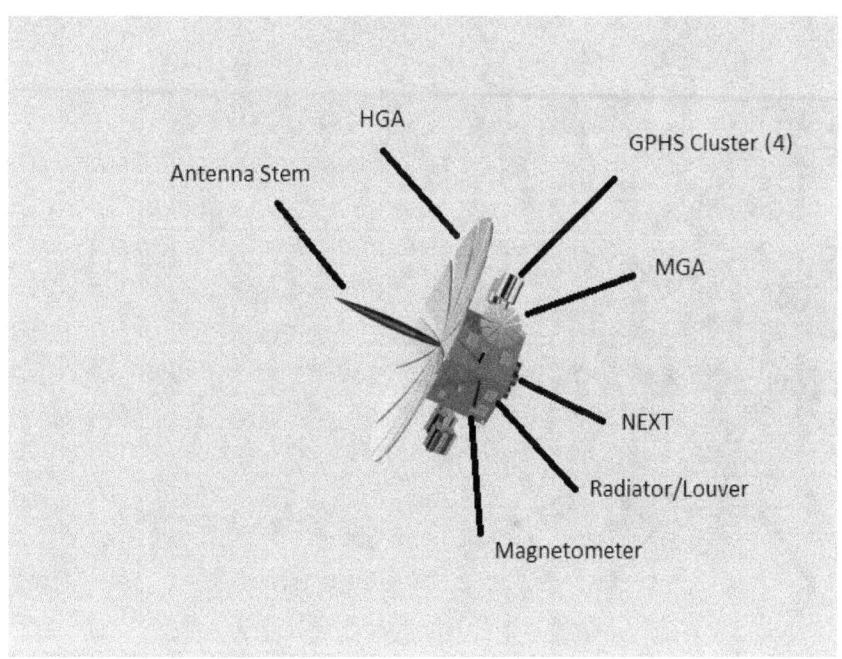

Figure 4.10: Zoomed Out View

- HGA/Radio Antenna in deployed configuration

- HGA/Radio Antenna is 12 meters in diameter

- Antenna Stem is 6 meters in height

- MGA is 2 meters in diameter

- Four next thrusters in a cluster

- Magnetometers on booms

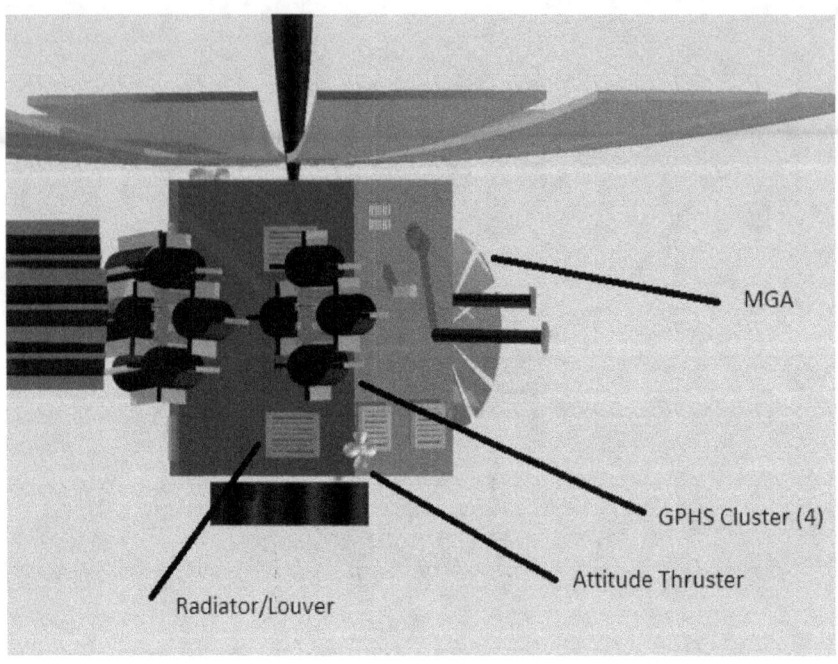

Figure 4.11: RTG View

- GPHS

    Four per boom

    Five booms on the spacecraft, for a total of twenty

- MLI

    Represented by the bronze exterior coloring

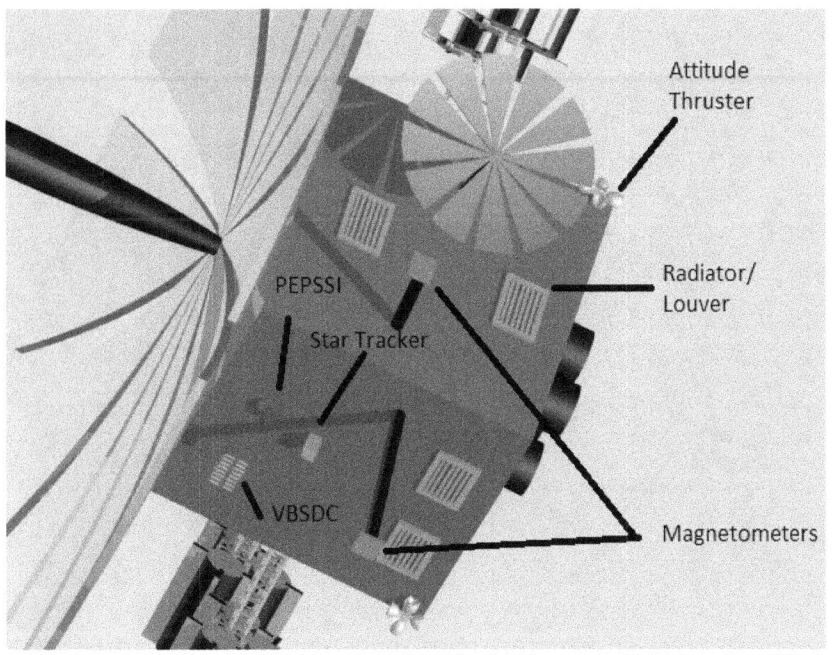

Figure 4.12: Zoomed In View

- Shows all secondary payloads

    Venetia Burney Student Dust Counter

    Flux Gate Magnetometer

    Scalar Helium Magnetometer

    Pluto Energetic Particle Spectrometer Science Investigation

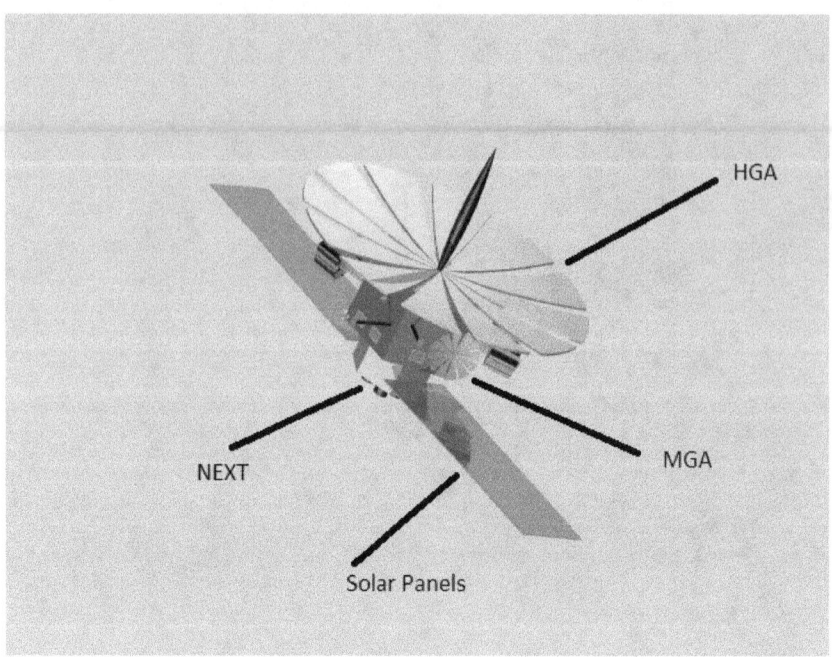

Figure 4.13: Spacecraft with Boost Phase

- Boost Phase Configuration

- Solar Panels

    3×10 m each

- Thrusters

    These four thrusters bring the total to eight

4.5 Performance Analysis

4.5.1 Mass. Table 4.10 accounts for the masses present on the spacecraft. If an exact number was not known a 25% margin was added to a best guess. Thus, the total mass of the spacecraft minus the weight of NEXT propellant is 2345.93 kg. The total mass of the boost phase minus the NEXT propellant is approximately 1350 kg and the mass of the NEXT xenon propellant is 4885 kg bringing the total launch mass to *8580.93 kg*.

Table 4.10: Spacecraft Dry Mass Budget

	Mass (kg)	Margin	Section
Attitude Control Propellant Tanks	2.66	.665	4.3.5.2
Attitude Control Propellant	150.72	37.68	4.3.5.2
Attitude Determination	6.89	-	
Xenon Propellant Tank	217.13	54.28	4.3.4
Cabling	20	5	-
MLI Blankets	41.61	10.40	4.3.6
RHU (10)	.397	-	2.4.4
Primary Payload and MGA	23.73	5.93	4.3.1
GPHS (20)	1156	-	4.3.3
Data Processing/Handling	15	3.75	-
Thrusters (4)	224.7	-	4.3.4
Louvers/Radiators (10)	11.96	-	4.3.6
Secondary Payloads	27.43	-	4.3.1
Radiation Vault	189	47.25	4.3.6
Structure	75	18.75	-
Total (with Margin)	2162.23(2345.93)		

4.5.2 Power. Table 4.11 presents the power draw of all of the spacecraft subsystems along with a 25% margin. The total power draw is well under the power provided by the RTGs displayed in Figure 4.14. In the figure the spike represents the post-Jupiter thrust profile which will be discussed in the following section. Furthermore, it is also clear from the figure that the spacecraft should provide enough power to send data for over 200 years.

Table 4.11: Spacecraft Power Consumption

	Power (W)	Margin (W)	Section
Secondary Payloads	25.4	-	4.3.1
Communications	15	3.75	4.3.2
Attitude Thrusters	50	12.5	4.3.5.2
Attitude Sensors	31	7.75	4.3.5.1
Thermal	20	5	4.3.6
Data Processing/Handling	15	3.75	4.3.2
Total (w/ Margin)	156 (188.75)		

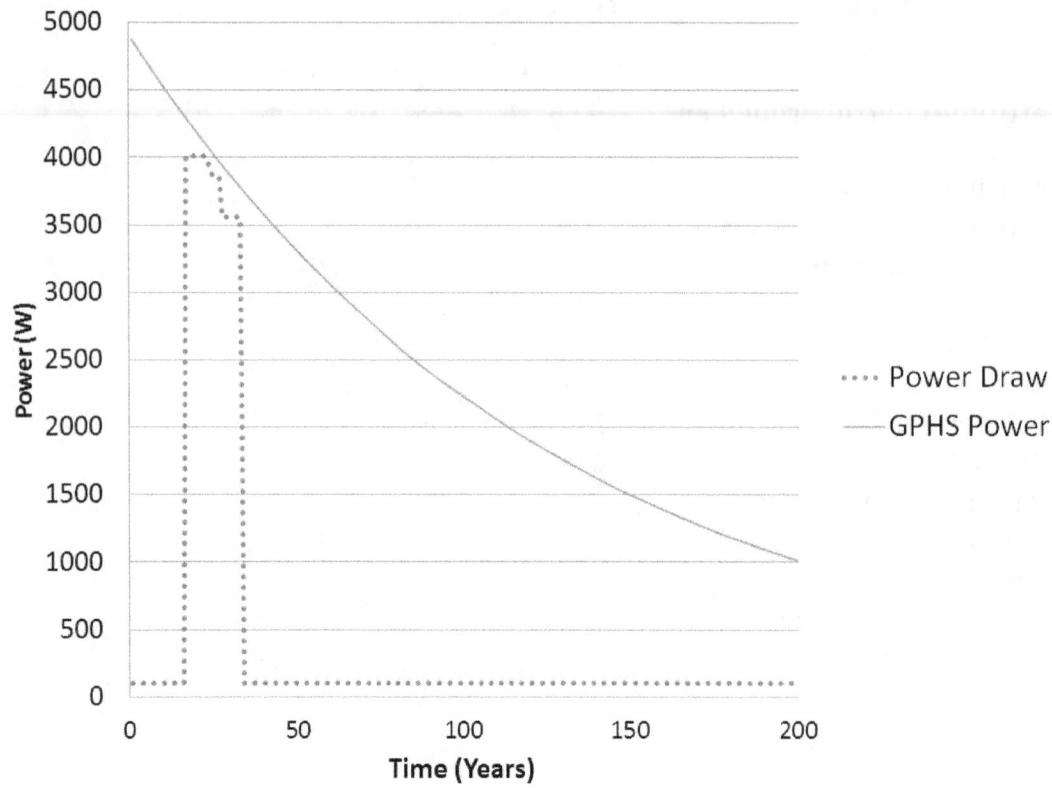

Figure 4.14: RTG Power Level and Spacecraft Power Draw

4.5.3 Trajectory Analysis. The trajectory analysis was split into three subsections to facilitate the analysis. The first subsection is the Pre-Jupiter Phase. This phase commences at the end of the launch vehicle insertion. Thus, at the start of this phase the spacecraft has the same velocity relative to the Sun as Earth–29.78 *km/sec*. Furthermore, this phase lasts until the Jupiter gravity assist at which point the Jupiter Phase begins using the final values from the first phase as its initial conditions. The Jupiter Phase consists solely of the gravity assist. Finally, the last phase is the Post-Jupiter Phase. This phase lasts from the termination of the gravity assist until the spacecraft reaches 550 AU where it will begin to examine the gravitational lens.

4.5.3.1   Pre-Jupiter Phase.   The spacecraft trajectory starts at Earth. Further, it is assumed that Earth and Jupiter are properly aligned for rendezvous. The initial angular position of Earth relative to the Sun is taken as the initial angle of the trajectory. Meaning the trajectory will begin at zero degrees. The complete trajectory from the Earth to Jupiter is shown in Figure 4.15. Since the trajectory is using low thrust propulsion the spacecraft naturally spirals around the body it is orbiting, and in this case that body is the sun.

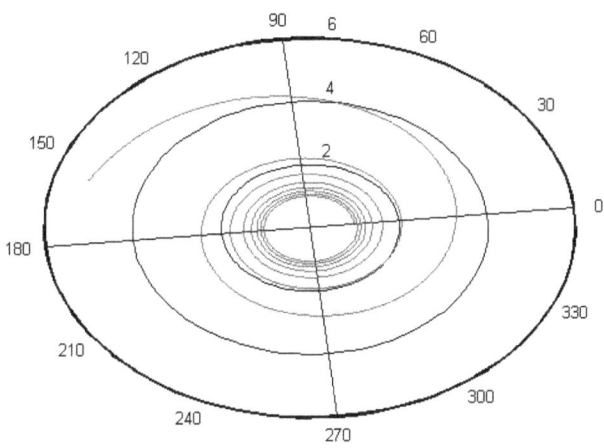

Figure 4.15: Pre-Jupiter Trajectory (Distance - AU)(Angle - degrees)

Travelling from the Earth to Jupiter on this trajectory requires 16.9 years, and during those years the spacecraft is thrusting the entire time. It is important to note that while the spacecraft is thrusting the entire time it is also being throttled to lower powers to compensate for the dwindling solar flux. The code in Appendix A.1 shows how this was simulated.

This trajectory requires 17.17 km/sec of ΔV, which corresponds to a propulsive mass of 2996.2 kg. Hence, the final spacecraft mass for this phase is 5603.8 kg. Finally, the spacecraft velocity at Jupiter is 13.68 *km/sec*. The entire velocity profile from Earth to Jupiter is shown in Figure 4.16. In the figure the velocity of the spacecraft is on a constant decline due to the pull of the sun's gravitation. However, the thrust of the spacecraft decreases the deceleration effect of the sun allowing the spacecraft to push itself all the way to Jupiter. Finally, the jagged portion of the profile is the result of the spacecraft's varying thrust levels.

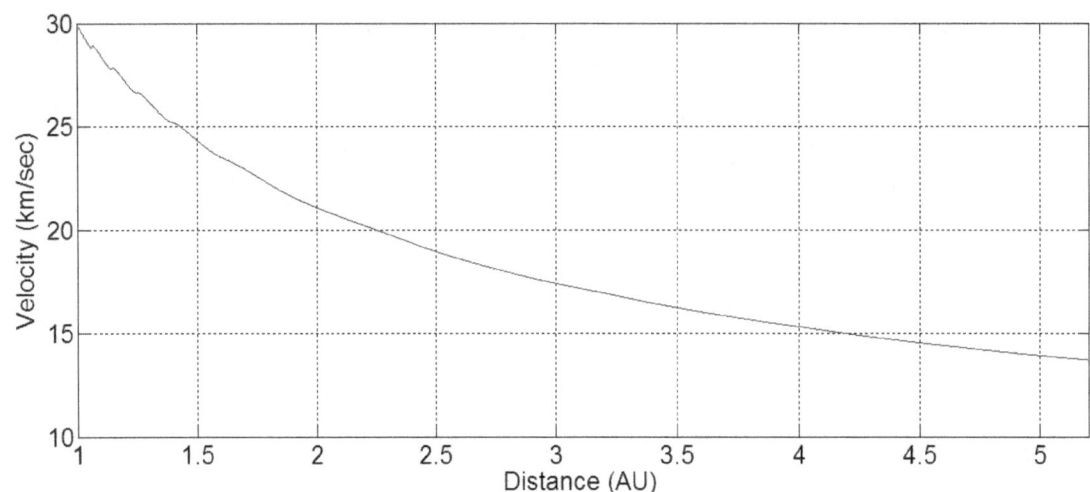

Figure 4.16: Pre-Jupiter Velocity Profile

4.5.3.2 Jupiter Phase. At Jupiter, the spacecraft enters into its gravity assist maneuver and drops its 1200 *kg* boost stage at Jupiter's periapsis. As a result, at the end of the gravity assist maneuver the spacecraft mass is down to 4403.8 *kg*. All of the gravity assist data was derived from the equations presented in Section 3.5.2. As expected, the velocities of the spacecraft relative to Jupiter entering and exiting the maneuver are equal, but the velocities relative to the Sun are distinctly different. In fact, relative to the Sun the spacecraft experiences a velocity increase of 13.905 *km/sec* bringing its exit velocity to 27.585 *km/sec*.

91

4.5.3.3 Post-Jupiter Phase.   The final phase characterizes the trajectory from Jupiter to 550 AU. The initial velocity at Jupiter is 27.585 *km/sec*. By the time the spacecraft reaches 550 AU its velocity has increased to 29.6 *km/sec*, or 6.24 *AU/year*. Figure 4.17 presents the complete velocity profile.

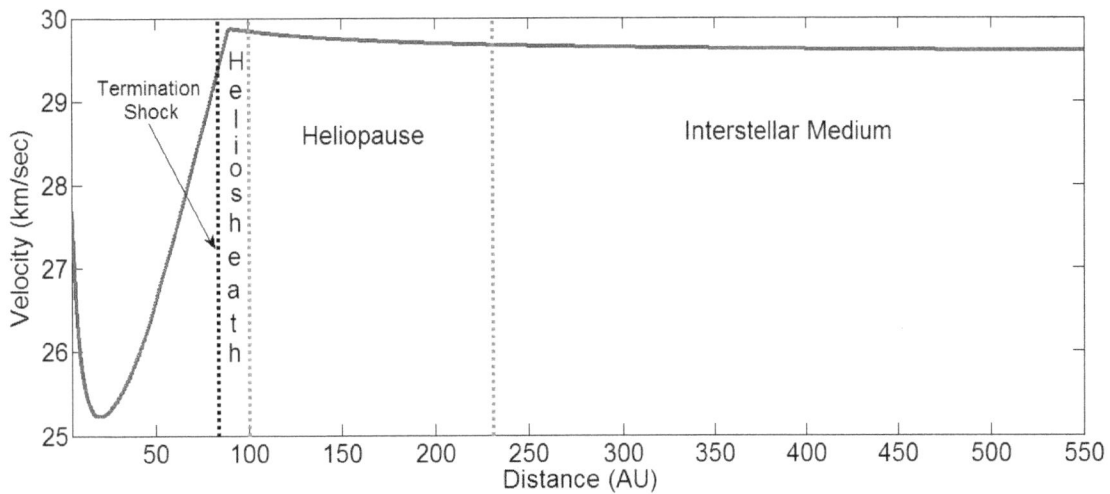

Figure 4.17: Post-Jupiter Velocity Profile

Looking at the figure, the velocity of the spacecraft continues to decline until a minimum velocity is reached. This is due to the waning effect of the sun's gravitation and the constant, long duration thrust of the propulsion system. Prior to the minimum the sun's gravitational force is dominant and decelerates the spacecraft. After the minimum the thrust of the propulsion system dominates, accelerating the spacecraft back to approximately 29.6 *km/sec*. Moreover, the velocity profile corresponds to a ΔV of 20.72 *km/sec* and a thrust period of 16.79 years.

4.5.3.4 Total.   These three phases result in a total ΔV of 51.79 *km/sec*, total thrust period of 33.69 years, and total trip time of 108.72 years. This may appear to be a

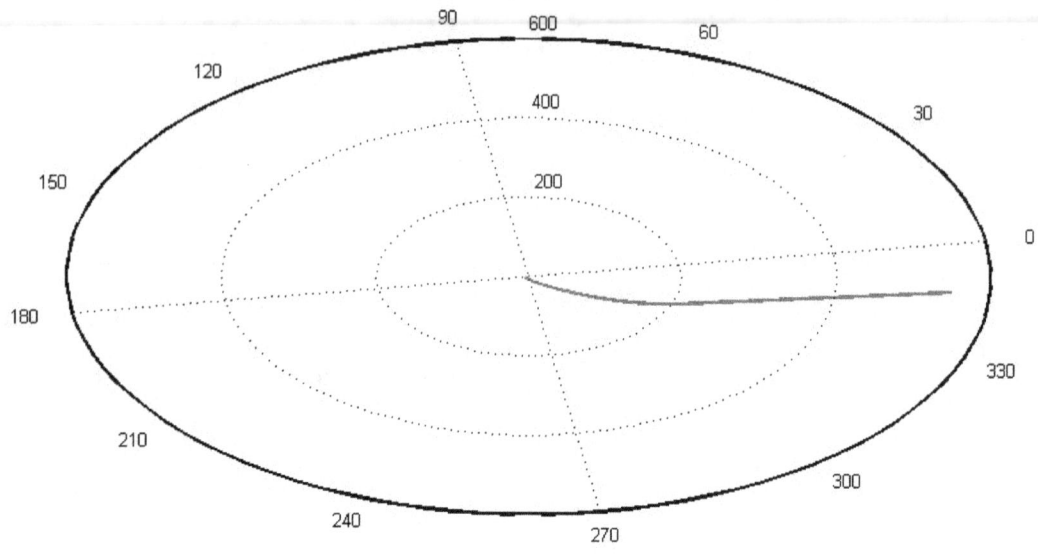

Figure 4.18: Post-Jupiter Trajectory (Distance - AU)(Angle - degrees)

daunting trip time but compare that to Voyager I. It was launched 35 years ago in 1977 and it would still take Voyager I another 119 years to reach 550 AU at 3.599 AU/year [48].

4.5.4 Communications. Figure 4.19 depicts the data rate for the mission assuming the communications architecture is receiving a full 15 W for transmit. That transmitter power corresponds to an *EIRP* of 15225.19 *W* (41.82 *dB*) for the MGA and 548106.7 *W* (57.39 *dB*) for the HGA. Moreover, the data rate in Figure 4.19 was calculated using Equation 3.10 with a spacecraft BER of 10^{-5}, BPSK, and Plus RS Viterbi Decoding. Finally, Figure 4.19 show the link requirement. If the HGA and MGA are above the minimum there is link margin. On the other hand, if it dips below the minimum it is no longer meeting the data transfer requirements. Therefore, at approximately 250 AU the MGA no longer meets the data transfer rate. This was expected and by this time the HGA will be handling data transfer.

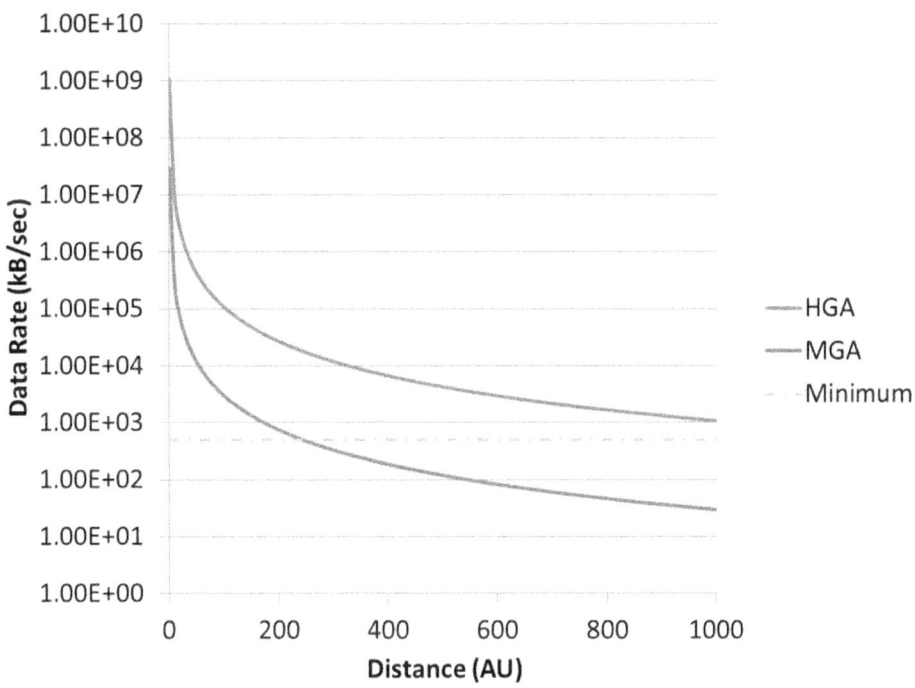

Figure 4.19: Mission Data Rate

4.5.5 Cost. Cost can be estimated a number of ways. One approach is to utilize cost models and that is what was done in this thesis. Two models were examined. The first model investigated was the NASA Advanced Missions Cost Model (AMCM). Its self-described purpose is to "provide a useful method for quick turnaround, rough-order-of-magnitude [cost estimation]" [49]. The inputs to the model are quantity, dry weight, mission type, launch year, block number, and difficulty. For the FSM, the quantity is one, the dry weight is 2345 kg, the mission type is "spacecraft - planetary", the launch year was chosen to be 2020, the block number (which is described as the level of inheritance from previous spacecraft) is one since the FSM constitutes a new system, and the difficulty is very high [49]. The cost estimation equation used by the AMCM model is

$$Cost = a * Q^b * W^c * d^S * e^{\frac{1}{IOC-1900}} * B^f * g^D \ [50] \tag{4.4}$$

94

where Q is the number of spacecraft, W is the weight, s is a value derived from the mission type, IOC is the launch year, B is is block number, D is a value derived from the difficulty, and a–g are model parameters [50]. The resulting cost is $3.927 billion (in FY11 US$).

The total cost is expected to be in the range of $3-5 billion. An exact cost cannot be tied down due to cost overruns, inflation, and the nebulous costs of labor. Nonetheless, the pay of a single GS-14 employee was analyzed to get a general idea of these costs. In 2012, a GS-14 working in Denver, CO will make $103,771 [51]. In the 108 years that it will take the spacecraft to reach 550 AU and given a conservative 2% raise per year that will increase to $880,835. That means, assuming spacecraft were launched today with a complement of 20 GS-14 engineers, labor costs in 2120, when the spacecraft arrives, would be approximately $17 million per year. What is more, integrating the cost from the launch date to 108 years shows that the total cost over that time span of all twenty positions is approximately $800 million. Outside of labor costs there are also operating costs like electricity, building maintenance, and equipment replacement costs which are outside of the scope of this analysis.

5 Conclusions and Recommendations

5.1 Introduction

Outside of the deliverables this thesis had three goals. The first goal was to produce a piece of work with academic merit. The second goal was to present a novel design. The final goal was to present actionable recommendations to the scientific community. The first four chapters of this thesis accomplished the first two goals and this final chapter accomplishes the third goal by presenting the conclusions of this research and analyzing them to come up with actionable recommendations.

5.2 Conclusions

In simple terms, this thesis demonstrated that it is possible to create an interstellar precursor spacecraft solely from commercial of the shelf parts. Unfortunately, it would take more than a century for the spacecraft to reach the desired distances meaning there would be significant risk and astronomical costs involved. Also, that century would eclipse the span of several mission directors and engineers lifetimes, which would introduce even more risk. Furthermore, if this mission were to fly successfully in spite of the obstacles the total gain from the sun and the 12-m antenna would only be 115 dB. A terrestrial based 8.4 km radio telescope would have the same gain, and there are already telescopes of that size. The gravitational lens still boasts superior angular resolution, staring time, and a clear view of its target without atmospheric interference. Nonetheless, there is no need to explore this mission further at this point because of the costs and trip times involved. Still, this thesis provided insight into what the current deep space capabilities are, and anyone interested in deep space flight could glean helpful information from it.

5.3 Recommendations

- The first recommendation is to spend more effort developing spacecraft power instead of new thrusters. At this moment, power is the limiting factor in deep space exploration; thus, it would most significantly improve deep space flight. For example, if the power density was increased so that the NEXT thruster could maintain maximum thrust for the entire FSM mission the trip time would decrease by nearly 10% to 98 years. If two thrusters could be powered there would be a 25% decrease in trip time to 80 years. This is because at higher powers the thruster can maintain higher thrust and use the propellant mass more efficiently with higher specific impulses. Spacecraft power can be developed through the exploration of new radioisotopes and higher efficiency solar panels or the introduction of new power methods, i.e. nuclear reactors.

- Again this recommendation relates to power. Though the exact amount of plutonium-238 the government possesses is a secret, it is estimated that there may only be enough plutonium-238 for one more mission. [52] Hence, the growing difficulty in procuring radioisotopes necessitates developing a novel power method such as nuclear reactors or restarting radioisotope production if deep space exploration is to continue.

- In this thesis, the long term reliability of all instruments was taken on the fact that the instruments chosen are manufactured to withstand the radiation that the FSM will experience. Moreover, it would not be feasible to do a long duration test that lasts more than a decade because by that time the device will be out of date. Hence, a novel long duration test method must be developed.

- In regard to communications, new modulation schemes must continue to be explored to push data rates closer to their Shannon Limit.

97

- Increase emphasis on low thrust propulsion, investigate replacement propellants because of the difficulty in procuring xenon. Xenon is a rare gas and it is only produced at air separation plants. There are only about 75 of these plants in the entire world [53]. These 75 plants produce just over 9 million liters or 53,046 kg of xenon a year which means the FSM, and any similar mission, would require nearly 10% of the world's yearly supply of xenon [53]. Thus xenon is not a sustainable propellant option.

- The continued development of a heavy launch vehicle on the scale of NASA's Space Launch System would enable deep space probes to carry significantly more propellant and power which would in turn decrease the trip time as shown in the first bullet.